Understanding Function Spaces

Bilal Bilalov

Editor

Understanding Function Spaces

© Copyright 2025 by Nova Science Publishers, Inc.
https://doi.org/10.52305/XBUN6843

All rights reserved. . No part of this book may be reproduced, stored in a retrieval system or transmitted in any form or by any means: electronic, electrostatic, magnetic, tape, mechanical photocopying, recording or otherwise without the written permission of the Publisher.
We have partnered with Copyright Clearance Center to make it easy for you to obtain permissions to reuse content from this publication. Simply navigate to this publications page on Novas website and locate the Get Permission button below the title description. This button is linked directly to the titles permission page on copyright.com. Alternatively, you can visit copyright.com and search by title, ISBN, or ISSN.

For further questions about using the service on copyright.com, please contact:

Copyright Clearance Center

Phone: +1-(978) 750-8400 Fax: +1-(978) 750-4470 E-mail: info@copyright.com

NOTICE TO THE READER

The Publisher has taken reasonable care in the preparation of this book, but makes no expressed or implied warranty of any kind and assumes no responsibility for any errors or omissions. No liability is assumed for incidental or consequential damages in connection with or arising out of information contained in this book. The Publisher shall not be liable for any special, consequential, or exemplary damages resulting, in whole or in part, from the readers use of, or reliance upon, this material. Any parts of this book based on government reports are so indicated and copyright is claimed for those parts to the extent applicable to compilations of such works.

Independent verification should be sought for any data, advice or recommendations contained in this book. In addition, no responsibility is assumed by the Publisher for any injury and/or damage to persons or property arising from any methods, products, instructions, ideas or otherwise contained in this publication.

This publication is designed to provide accurate and authoritative information with regard to the subject matter covered herein. It is sold with the clear understanding that the Publisher is not engaged in rendering legal or any other professional services. If legal or any other expert assistance is required, the services of a competent person should be sought. FROM A DECLARATION OF PARTICIPANTS JOINTLY ADOPTED BY A COMMITTEE OF THE AMERICAN BAR ASSOCIATION AND A COMMITTEE OF PUBLISHERS.

Library of Congress Cataloging-in-Publication Data

ISBN: 979-8-89530-627-7 (Hardcover)

Published by Nova Science Publishers, Inc. † New York

Contents

Introduction

Recently, there has been increasing interest in Banach function spaces related to some problems of mechanics, mathematical physics, and pure mathematics. These spaces include classical spaces of continuous functions, Lebesgue spaces, as well as so-called non-standard spaces such as Lebesgue spaces with variable summability index, Morrey spaces, grand Lebesgue spaces, Orlicz spaces, Marcinkiewicz spaces, Lorentz spaces, etc. The reasons for the emergence of these spaces are different. For example, the Morrey space was introduced in the study of smoothness properties of solutions to elliptic equations, variable Lebesgue spaces are significant in applications to partial differential equations and variational integrals with non-standard growth conditions, and grand Lebesgue spaces were introduced in the search for the minimal restriction on the Jacobian of a mapping of a multidimensional domain, etc.
Compared to other areas of mathematics, harmonic analysis is more deeply engaged with these spaces. Significant results have been obtained concerning the boundedness of singular Calderon-Zygmund operators, the Riesz potential, and the embedding theorems in these spaces. These results make it possible to study the solvability of differential equations in Sobolev spaces generated by the norms of the above spaces. Numerous monographs and research works have been dedicated to these kinds of problems (see, for example, [1–13] and references therein).

This book is devoted to problems related to mathematics regarding Banach spaces of functions. A brief overview is given of the main properties, embedding results, and the most important generalizations of different function spaces, such as Morrey spaces, Morrey-Campanato spaces, Holder spaces. The criteria that ensure the infinite system of integral equations has a solution in the corresponding Banach sequence space are also determined. Some problems of approximation theory in variable exponent Lebesgue spaces are investigated. Additionally, a nonlocal problem with nonlinearity for a second-order nonlinear Fredholm integro-differential equation with a degenerate kernel is considered. The understanding of well-posed and ill-posed problems, as well as the developed methods for solving ill-posed applied problems in mathematics is also studied. Finally, two conflict-controlled objects, whose movements are inertial, are considered. The results obtained for the latter have theoretical and practical significance in control theory.

Chapter 1

Morrey-Campanato Spaces and Some Generalizations

L. Softova *
University of Salerno, Department of Mathematics, Fisciano, Italy

To my children:
my strength and my weakness!

Abstract

The theory of *Campanato-Morrey spaces,* along with their applications and generalizations, has evolved significantly over the last few decades. These spaces now play a fundamental role in modern Function and Harmonic Analysis, as well as in the regularity theory of Partial Differential Equations, alongside classical Hölder and Sobolev spaces.

In his landmark paper *"On the solutions of quasi-linear elliptic partial differential equations"* (see [48]), Morrey introduced a novel approach to studying the local Hölder regularity of solutions to second-order elliptic operators. This method involved estimating the growth

*Corresponding Author's Email: lsoftova@unisa.it

of the integral function $g(r) = \int_{\mathscr{B}_r} |Du(y)|^p \, dy$ in terms of the power of the ball radius, that is, Cr^λ with $\lambda > 0$. Although Morrey did not explicitly introduce function spaces, his paper is widely regarded as the foundation for the theory of *Morrey spaces* $L^{p,\lambda}$ and their numerous generalizations.

In this chapter, we present a brief overview of the key properties, selected embedding results, and the most significant generalizations of the aforementioned function spaces.

Keywords: Morrey-Campanato spaces, embedding theorems, Maximal operator, Maximal inequality, Weighted spaces, Muckenhoupt weight

AMS Subject Classification: 42-00, 42B25, 42B35, 43A85, 46-01, 46E30, 46E35.

1. Morrey Spaces

The spaces $L^{p,\lambda}$, named after Morrey, were introduced in the 1960s and subsequently studied independently by Campanato, Peetre, Spanne, and Brudnyi (see [18,21,22,48,57,71]). The development of these new classes of functions, together with their various generalizations and the associated embedding theory, has led to substantial advancements in the classical theory of Sobolev spaces.

1.1. Classical Morrey Spaces, Definitions and Main Properties

Let $\Omega \subset \mathbb{R}^n, n \geq 2$, be a bounded domain, i.e., a connected open set with finite diameter:

$$d_\Omega = \sup_{x,y \in \Omega} |x - y| < \infty.$$

For any ball $\mathscr{B}_r(x) \subset \mathbb{R}^n$, centered at $x \in \Omega$ with radius $r \in (0, d_\Omega]$, we denote

$$\Omega_r(x) = \Omega \cap \mathscr{B}_r(x).$$

Throughout this chapter, Ω is assumed to be either a domain of type (A) or one that satisfies the so-called *Campanato condition* (cf. [20–22]). Specifically, there exists a positive constant A_Ω such that

$$|\Omega_r(x)| \geq A_\Omega \, r^n \qquad \forall \, x \in \overline{\Omega}, \; r \in (0, d_\Omega]. \tag{A}$$

Definition 1.1. *A measurable function $f\colon \Omega \to \mathbb{R}$ belongs to the Morrey space $L^{p,\lambda}(\Omega)$, where $1 \le p < \infty$ and $0 \le \lambda < n$, if there exists a positive constant M such that*

$$\int_{\Omega_r(x)} |f(y)|^p \, dy \le M r^\lambda \qquad \forall\, x \in \Omega,\ r \in (0, d_\Omega].$$

As a norm in $L^{p,\lambda}(\Omega)$, we take the expression

$$\|f\|_{L^{p,\lambda}(\Omega)} = \sup_{x\in\Omega,\, r>0} \left(\frac{1}{r^\lambda} \int_{\Omega_r(x)} |f(y)|^p \, dy \right)^{1/p}. \tag{1.1}$$

By identifying functions that are equal almost everywhere with respect to the Lebesgue measure, this defines a normed space.

It is easy to see that (1.1) defines a homogeneous norm, meaning that for any $t > 0$, the following holds:

$$\|f(t\cdot)\|_{L^{p,\lambda}(\Omega)} = t^{\frac{\lambda-n}{p}} \|f\|_{L^{p,\lambda}(\Omega)}.$$

Theorem 1.1. *The Morrey spaces are Banach spaces.*

Proof. Let $\{f_j\}$ be a Cauchy sequence in $L^{p,\lambda}(\Omega)$. Then, there exists a positive constant K such that $\|f_j\|_{L^{p,\lambda}(\Omega)} \le K$ for every $j \in \mathbb{N}$. Since

$$\|f\|_{L^p(\Omega)} \le d_\Omega^{\lambda/p} \|f\|_{L^{p,\lambda}(\Omega)} \qquad \forall\, f \in L^{p,\lambda}(\Omega),$$

it follows that $\{f_j\}$ is also a Cauchy sequence in $L^p(\Omega)$. The completeness of $L^p(\Omega)$ garanties that $\{f_j\}$ converges with respect to the $L^p(\Omega)$-norm to some $f \in L^p(\Omega)$.

Thus, for almost every $x \in \Omega$ and every $r \in (0, d_\Omega]$, we have

$$\begin{aligned}
\frac{1}{r^{\lambda/p}} \|f\|_{L^p(\Omega_r(x))} &\le \frac{1}{r^{\lambda/p}} \|f - f_j\|_{L^p(\Omega)} + \frac{1}{r^{\lambda/p}} \|f_j\|_{L^p(\Omega_r(x))} \\
&\le \frac{1}{r^{\lambda/p}} \|f - f_j\|_{L^p(\Omega)} + \|f_j\|_{L^p(\Omega_r(x))} \\
&\le \frac{1}{r^{\lambda/p}} \|f - f_j\|_{L^p(\Omega)} + K,
\end{aligned}$$

which converges to K as $j \to \infty$. This implies that $f \in L^{p,\lambda}(\Omega)$, so the space is complete.

To show that $\|f - f_j\|_{L^{p,\lambda}(\Omega)} \to 0$ as $j \to \infty$, we fix $\varepsilon > 0$. Then there exists an index $N(\varepsilon) \in \mathbb{N}$ such that $\|f_j - f_k\|_{L^{p,\lambda}(\Omega)} < \varepsilon$ for all $j,k > N(\varepsilon)$. For almost all $x \in \Omega$ and all $r \in (0, d_\Omega]$, we have

$$\begin{aligned}\frac{1}{r^{\lambda/p}}\|f-f_j\|_{L^p(\Omega_r(x))} &\le \frac{1}{r^{\lambda/p}}\|f-f_k\|_{L^p(\Omega_r(x))} + \frac{1}{r^{\lambda/p}}\|f_k-f_j\|_{L^p(\Omega_r(x))}\\ &\le \frac{1}{r^{\lambda/p}}\|f-f_k\|_{L^p(\Omega)} + \|f_j-f_k\|_{L^{p,\lambda}(\Omega)}\\ &< \frac{1}{r^{\lambda/p}}\|f-f_k\|_{L^p(\Omega)} + \varepsilon, \qquad \forall\, j,k > N(\varepsilon).\end{aligned}$$

Passing to the limit as $k \to \infty$ and using (1.1), we obtain the desired result. □

Another important property of the Morrey norm is the validity of the *convolution inequality:*

$$\|f * g\|_{L^{p,\lambda}(\Omega)} \le \|g\|_{L^1(\Omega)}\|f\|_{L^{p,\lambda}(\Omega)}, \qquad 0 \le \lambda < n, \quad 1 \le p < \infty$$

for every $g \in L^1(\Omega)$ and $f \in L^{p,\lambda}(\Omega)$.

The definition of Morrey spaces can also be extended in various ways to include functions defined locally on the whole $\mathbb{R}^n$ (see for instance [59] and the references therein).

Definition 1.2. *We say that $f \in L^{p,\lambda}(\mathbb{R}^n)$ with $1 \le p < \infty$ and $0 \le \lambda < n$ if $f \in L^p_{\mathrm{loc}}(\mathbb{R}^n)$ and*

$$\|f\|_{L^{p,\lambda}(\mathbb{R}^n)} = \sup_{x\in\mathbb{R}^n,\, r>0}\left(\frac{1}{r^\lambda}\int_{\mathscr{B}_r(x)}|f(y)|^p\,dy\right)^{1/p} < \infty. \tag{1.2}$$

Let $x_0 \in \mathbb{R}^n$. The local Morrey space $L^{p,\lambda}_{\mathrm{loc}}(\mathbb{R}^n)$ consists of all functions $f \in L^p_{\mathrm{loc}}(\mathbb{R}^n)$ for which the following norm is finite:

$$\|f\|_{L^{p,\lambda}_{\mathrm{loc}}(\mathbb{R}^n)} = \sup_{r>0}\left(\frac{1}{r^\lambda}\int_{\mathscr{B}_r(x_0)}|f(y)|^p\,dy\right)^{1/p} < \infty. \tag{1.3}$$

1.2. Embedding Properties of the Morrey Spaces

The spaces $L^{p,\lambda}$ are defined for $\lambda \in (0,n)$. This restriction is natural, as can be observed from the following inclusions: for all $1 \le p < \infty$ and Ω satisfying the (A)-condition:

1. $L^{p,0}(\Omega) \cong L^p(\Omega)$.
2. $L^{p,n}(\Omega) \cong L^\infty(\Omega)$, by the Lebesgue point arguments.

3. $L^{p,\lambda}(\Omega) = \{0\}$ if $\lambda > n$.

   In fact, for $f \in L^{p,\lambda}(\Omega)$ and almost every $x \in \Omega$ we have

$$\frac{1}{r^\lambda}\int_{\Omega_r(x)} |f(y)|^p\, dy \leq \frac{1}{r^n}\int_{\Omega_r(x)} |f(y)|^p\, dy \leq Cr^{\lambda-n} \to 0 \quad \text{if } r \to 0$$

   and by the Lebesgue point argument, we conclude that $f(x) = 0$ for almost every $x \in \Omega$.

4. $L^{p,\lambda}(\Omega) = \{0\}$ if $\lambda < 0$.

   Obviously $\sup_{r>0} r^{-\lambda} \|f\|^p_{p,\mathscr{B}_r} < C$ implies $f(x) = 0$ for almost all $x \in \Omega$.

   Note that the above inclusions also hold for spaces defined on $\mathbb{R}^n$.

5. The family of the Morrey spaces $L^{p,\lambda}(\Omega)$ is partially ordered (see [20, 58]), in the sense that for each $1 \leq p \leq q < \infty$ and $\lambda, \mu \in [0,n)$ the following embedding holds:

$$L^{q,\mu}(\Omega) \hookrightarrow L^{p,\lambda}(\Omega) \qquad \text{if} \quad \frac{n-\lambda}{p} \geq \frac{n-\mu}{q}. \tag{1.4}$$

   This inclusion is also valid in $\mathbb{R}^n$ but under the condition $q(n-\lambda) = p(n-\mu)$.

Let $f \in L^{p,\lambda}(\Omega)$ and $g \in L^{q,\mu}(\Omega)$. Direct calculations, based on the classical Hölder inequality applied to integrals over balls, lead to a version of the Hölder inequality in Morrey spaces. More precisely, for any $p, q \in [1,\infty)$, such that $\frac{1}{p} + \frac{1}{q} \geq 1$ we have

$$\|fg\|_{L^{r,\nu}(\Omega)} \leq \|f\|_{L^{p,\lambda}(\Omega)} \|g\|_{L^{q,\mu}(\Omega)} \tag{1.5}$$

provided that

$$\frac{1}{r} = \frac{1}{p} + \frac{1}{q}, \qquad \frac{\nu}{r} = \frac{\lambda}{p} + \frac{\mu}{q}.$$

The Sobolev-Morrey spaces $W_k^{p,\lambda}(\Omega)$, $k \in \mathbb{N}$, built upon $L^{p,\lambda}(\Omega)$, $p \geq 1, \lambda \in [0,n)$, consist of all Morrey functions defined on Ω that have distributional derivatives up to order k, with these derivatives also belonging to the same Morrey space. The norm in $W_k^{p,\lambda}(\Omega)$ is defined analogously to the norm of the classical Sobolev spaces, i.e.

$$\|f\|_{W_k^{p,\lambda}(\Omega)} = \sum_{|\alpha|=0}^{k} \|D^\alpha f\|_{L^{p,\lambda}(\Omega)}$$

where α is a multi-index, $|\alpha| = \sum_{i=1}^{n} \alpha_i$, and $D^\alpha f = D_{x_1}^{\alpha_1} \dots D_{x_n}^{\alpha_n} f$. The following analogue of the Sobolev embedding theorem is proved by Campanato in [20] (see also [59]).

Denote by p_λ^* the Sobolev-Morrey conjugate of p, that is,

$$p_\lambda^* = \begin{cases} \dfrac{p(n-\lambda)}{n-\lambda-kp} & \text{if } 0 \le \lambda < n-kp \\ +\infty & \text{if } \lambda = n-kp \end{cases}$$

for each $k \in \mathbb{N}$.

Theorem 1.2 ([20]). *Let Ω be of type* (A) *and $f \in W_k^{p,\lambda}(\Omega)$, where $p \in [1,\infty)$ and $\lambda \in (0,n)$.*

- *If $0 \le \lambda \le n-kp$, then $f \in L^{q,\mu}(\Omega)$ for each $1 \le q < p_\lambda^*$, $\mu < n + \dfrac{q(kp+\lambda-n)}{p}$, and the following estimate holds:*

$$\|f\|_{L^{q,\mu}(\Omega)} \le c\|f\|_{W_k^{p,\lambda}(\Omega)}.$$

- *If $n-lp < \lambda \le n$ for some $1 \le l \le k$, then $f \in C^{k-l}(\overline{\Omega})$ and*

$$\sum_{|\beta|=k-l} \max_{x\in\overline{\Omega}} |D^\beta f(x)| \le c\|f\|_{W_k^{p,\lambda}(\Omega)}.$$

Theorem 1.2 still holds even if we assume that $f \in W_k^p(\Omega)$ and only the higher order derivatives of f belong to $L^{p,\lambda}(\Omega)$.

The importance of the embedding properties of Morrey spaces is closely related to the regularity theory for PDEs. One significant application of Theorem 1.2 is the analysis of the precise regularity of solutions to higher-order uniformly elliptic equations. In the paper [56], we provide a characterization of the Morrey, *BMO* and Hölder regularity of functions belonging to fixed Sobolev-Morrey spaces.

Let us recall that the class of functions with *Bounded Mean Oscillation* (*BMO*) and their subclass of functions with *Vanishing Mean Oscillation* (*VMO*) were introduced by John-Nirenberg in [40] and Sarason in [62], respectively. For completeness, we provide the definitions of these classes of discontinuous functions.

Definition 1.3. *Let $a \in L^1_{\text{loc}}(\mathbb{R}^n)$ and let $a_{\mathscr{B}_r} = \dfrac{1}{|\mathscr{B}_r|}\displaystyle\int_{\mathscr{B}_r} a(y)\,dy$ be the mean integral of a. For $R > 0$, we denote by $\gamma_a(R)$ the quantity*

$$\gamma_a(R) = \sup_{\mathscr{B}_r, r \le R} \frac{1}{|\mathscr{B}_r|} \int_{\mathscr{B}_r} |a(y) - a_{\mathscr{B}_r}|\,dy.$$

We say that

- *$a \in BMO$ if* $\quad \|a\|_* = \sup_{R>0} \gamma_a(R) < +\infty.$

  The quantity $\|a\|_*$ defines a norm in BMO (modulo the null function), making BMO a Banach space.

- *$a \in VMO$ if $a \in BMO$ and*

  $$\lim_{R \to 0} \gamma_a(R) = 0.$$

  We refer to $\gamma_a(R)$ as the VMO-modulus of a.

For any bounded domain $\Omega \subset \mathbb{R}^n$, the spaces $BMO(\Omega)$ and $VMO(\Omega)$ are defined by taking $a \in L^1(\Omega)$ and considering the integral in $\gamma_a(R)$ over Ω_r instead of $\mathscr{B}_r$.

Any function $f \in BMO(\Omega)$ or $VMO(\Omega)$, defined over a bounded domain $\Omega \subset \mathbb{R}^n$ with the extension property, can be extended to a BMO or VMO function in the whole space, respectively, while preserving its BMO-norm or VMO-modulus (see [14, 41]).

The following result provides precise regularity for the intermediate derivatives of functions with higher-order Morrey integrability.

Theorem 1.3 ([56, 67]). *Let $f \in W^{p,\lambda}_{2b}(\Omega)$ and let s_0 be the smallest non-negative integer such that $\dfrac{n}{2b - s_0} > 1$. Then, for any $s \in \{s_0, \ldots, 2b-1\}$, we have:*

a) If $p \in \left(1, \dfrac{n-\lambda}{2b-s}\right)$, then

$$D^s f \in L^{p,(2b-s)p+\lambda}(\Omega).$$

b) If $p = \dfrac{n-\lambda}{2b-s}$, then $D^s f \in BMO(\Omega)$.

c) If $p \in \left(\frac{n-\lambda}{2b-s}, \frac{n-\lambda}{2b-s-1}\right)^{\dagger}$*, then*

$$D^s f \in C^{0,\sigma_s}(\Omega) \quad \text{with} \quad \sigma_s = 2b - s - \frac{n-\lambda}{p}.$$

d) If $s_0 \geq 1$ *(i.e.,* $2b \geq n$*) and* $p \in \left(1, \frac{n-\lambda}{2b-s_0}\right)$*, then*

$$f \in C^{s_0-1, 2b-s_0+1-\frac{n-\lambda}{p}}(\Omega).$$

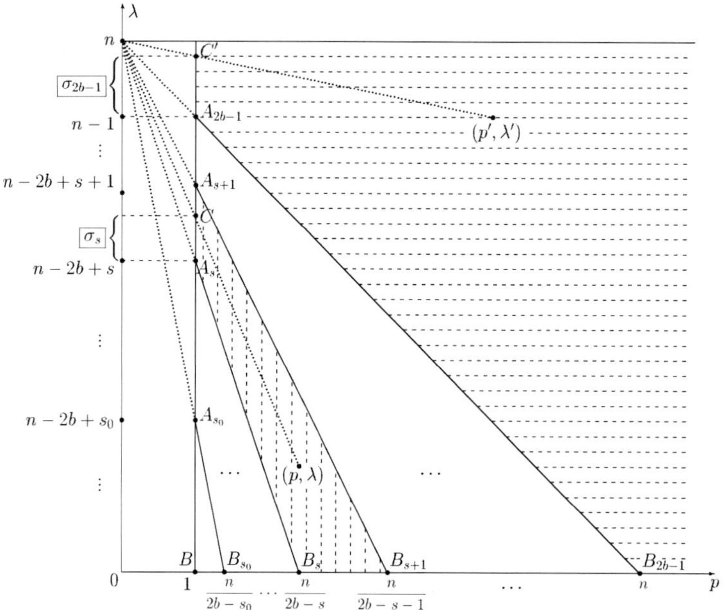

Figure 1.1. The plane $O_{p,\lambda}$

Theorem 1.3 provides a detailed characterization of the regularity of functions in Sobolev-Morrey spaces $W^{p,\lambda}_{2b}(\Omega)$ based on the values of p and λ. The geometric interpretation of the theorem, shown in Figure 1.1, is depicted in the half-strip $\{(p,\lambda) : p > 1, 0 < \lambda < n\}$ in the $p\lambda$-plane.

For each fixed $s \in \{s_0, \ldots, 2b-1\}$, the corresponding geometric region provides the boundaries for the regularity properties of the derivatives $D^s f$.

†This rewrites as $p \in (n-\lambda, \infty)$ when $s = 2b-1$.

- **Point** B_s: The coordinates of B_s are $B_s = \left(\frac{n}{2b-s}, 0\right)$, which corresponds to the threshold for p where the regularity conditions change.
- **Point** B: This is the point at $B = (1,0)$ on the p-axis, serving as a reference point.
- **Point** A_s: This point is the intersection of the vertical line $p = 1$ and the segment connecting $(0,n)$ and B_s. The coordinates of A_s are $A_s = (1, n-2b+s)$, marking the boundary between different Sobolev-Morrey regularity conditions.

The points A_s and B_s define the geometric region where the regularity of the derivatives $D^s f$ transitions based on the values of p, λ, and s.

a) For points (p,λ) lying in the triangle $\triangle BB_sA_s$ we obtain specific regularity results for the derivatives of f:

$$D^s f \in L^{p,(2b-s)p+\lambda}(\Omega).$$

In particular, for $(p,\lambda) \in \triangle BB_{s_0}A_{s_0}$, we have the regularity result for

$$D^{s_0} f \in L^{p,(2b-s_0)p+\lambda}(\Omega) \quad \text{and} \quad f \in C^{s_0-1,2b-s_0+1-\frac{n-\lambda}{p}}(\Omega)$$

if $s_0 \geq 1$.

b) If (p,λ) lies on the segment A_sB_s, we have $D^s f \in BMO$. In particular, if $(p,\lambda) \in A_{s_0}B_{s_0}$, then we have $D^{s_0} f \in BMO$.

c) Let $s \in \{s_0, \ldots, 2b-2\}$, and suppose (p,λ) lies in the interior of the quadrilateral $Q_s := B_sB_{s+1}A_{s+1}A_s$. Then $D^s f \in C^{0,\sigma_s}(\Omega)$ whereas

$$D^{s+1} f \in L^{p,(2b-s-1)p+\lambda}(\Omega).$$

Moreover, the exponent σ_s is the length of the segment CA_s, where $C = C(p,\lambda)$ is the intersection of the line $p = 1$ with the line passing through the points (p,λ) and $(0,n)$. In particular, $(p,\lambda) \in A_{s+1}B_{s+1}$ implies that $D^{s+1} f \in BMO$.

Similarly, set Q_{2b-1} for the shadowed *unbounded region* in Figure 1.1. Then, if $(p',\lambda') \in Q_{2b-1}$, we have $D^{2b-1} f \in C^{0,\sigma_{2b-1}}(\Omega)$, with σ_{2b-1} equal to the length of $C'A_{2b-1}$, where $C' = C'(p',\lambda')$, while $D^{2b-1} u \in BMO$ if $(p',\lambda') \in A_{2b-1}B_{2b-1}$.

Analogous embedding results have also been obtained for parabolic Sobolev-Morrey spaces, describing the regularity properties of higher-order parabolic operators acting in these function classes; see [55, 67].

One of the most important results, concerning the Morrey spaces is the validity of the *Maximal inequality.* As is well known, the *fundamental Lebesgue theorem* asserts that

$$\lim_{r\to 0}\frac{1}{|\mathscr{B}_r(x)|}\int_{\mathscr{B}_r(x)} f(y)\,dy = f(x) \qquad \text{for a.a. } x\in\mathbb{R}^n \tag{1.6}$$

where f is a locally integrable function defined on $\mathbb{R}^n$. To study this limit, its quantitative analogue

$$\mathscr{M}f(x) = \sup_{x\in\mathbb{R}^n, r>0}\frac{1}{|\mathscr{B}_r(x)|}\int_{\mathscr{B}_r(x)} |f(y)|\,dy$$

called the *Maximal operator* is very useful. It is readily seen that $\mathscr{M}f$ is a positive, measurable, lower semicontinuous function, but it is not integrable, as the following example shows (cf. [73]).

Example 1.4. *Take*

$$f(y) = \chi_{(0,1/2)}(y)\frac{d}{dy}\frac{1}{\ln(1/y)}.$$

Then $f(y) > 0$ and for $x\in(-1/2,0)$ we have

$$\mathscr{M}f(x) \ge \frac{1}{2r}\int_{x-r}^{x+r} f(y)\,dy \ge \frac{1}{4|x|}\frac{1}{\ln(1/|x|)}$$

which is not integrable in a neighbourhood of 0. □

To overcome this inconvenience, the class of integrable functions proves to be of great help. A fundamental result concerning $\mathscr{M}f$ and related to the works of *Hardy, Littlewood,* and *Stein* (see [73, 75]), is the following.

Theorem 1.5 (Maximal Inequality). *Let $f\in L^p(\mathbb{R}^n)$ with $1\le p\le\infty$. Then $\mathscr{M}f$ is finite almost everywhere, and the following results hold:*

1. *If $f\in L^1(\mathbb{R}^n)$, then for every $\tau > 0$*

$$|\{x : \mathscr{M}f(x) > \tau\}| \le \frac{c(n)}{\tau}\int_{\mathbb{R}^n} |f(y)|\,dy.$$

2. *If $f \in L^p(\mathbb{R}^n)$, with $1 < p \leq \infty$, then $\mathcal{M} f \in L^p(\mathbb{R}^n)$ and*

$$\|\mathcal{M} f\|_p \leq c(n,p)\,\|f\|_p.$$

Corollary 1.6. *If $f \in L^p(\mathbb{R}^n)$, $1 \leq p \leq \infty$ then* (1.6) *holds for almost every $x \in \mathbb{R}^n$.*

A natural question that arises is: *whether Theorem 1.5 still holds for functions measurable concerning a weighted measure $w(x)dx$ with a non-negative measurable weight w?*

The answer was provided by Muckenhoupt, who proved that if f belongs to the *weighted Lebesgue space* $L^{p,w}(\mathbb{R}^n)$, where $w\colon \mathbb{R}^n \to [0,\infty)$ is a weight, endowed with the norm

$$\|f\|_{p,w} = \int_{\mathbb{R}^n} |f(x)|^p w(x)\,dx < \infty,$$

then Theorem 1.5 remains true if and only if w satisfies the Muckenhoupt condition

$$\left(\frac{1}{|\mathscr{B}|}\int_{\mathscr{B}} w(x)\,dx\right)\left(\frac{1}{|\mathscr{B}|}\int_{\mathscr{B}} w(x)^{1/(1-p)}\,dx\right)^{p-1} < \infty \tag{1.7}$$

for any ball $\mathscr{B} \subset \mathbb{R}^n$. Theorem 1.5 is also valid in the setting of Morrey spaces, as shown by *Chiarenza* and *Frasca* in [25]. In addition to the maximal inequality, they also prove the boundedness of the *Riesz potential* and *Calderón-Zygmund singular integrals* acting on $L^{p,\lambda}$.

Recently there have been many extensions of the results of Chiarenza and Frasca in various spaces generalizing the growth function of the Morrey and weighted Morrey spaces (see for example [34,35,37,42,51,53,59,66] and the references therein).

Another operator that plays an important role in the study of the behavior of commutators of singular integrals and essentially bounded functions in various spaces is the *Sharp maximal operator* $f^\sharp$ defined as follows:

$$f^\sharp(x) := \sup_{\mathscr{B}_r(x)} \frac{1}{|\mathscr{B}_r(x)|}\int_{\mathscr{B}_r(x)} |f(y) - f_{\mathscr{B}_r(x)}|\,dy$$

where the supremum is taken over all balls $\mathscr{B}_r(x)$ in $\mathbb{R}^n$.

The Fefferman-Stein inequality (cf. [33, Theorem IV.2.20]) provides an estimate for the weighted Lebesgue norm $L^{p,w}(\mathbb{R}^n)$ of the maximal operator via the corresponding norm of the sharp operator. This estimate allows for the evaluation of the Morrey norm of functions through their Sharp maximal function (see [30,66] and the references therein).

Precisely, if $f \in L^{p,\lambda}(\mathbb{R}^n)$ with $p \in (1,\infty)$, $\lambda \in (0,n)$, then the following estimate holds:

$$\|f\|_{L^{p,\lambda}(\mathbb{R}^n)} \le c\,\|f^{\#}\|_{L^{p,\lambda}(\mathbb{R}^n)}. \tag{1.8}$$

The estimate (1.8) continuous to hold if we replace $\mathbb{R}^n$ with a bounded domain Ω of (A)-type (see [25, 30, 66]). It is enough to extend the function f as zero in $\mathbb{R}^n \setminus \Omega$ and then apply (1.8).

1.3. Generalized Morrey Spaces and Maximal Inequality

The first generalization of the Morrey spaces is given by *Spanne* in [71, 72] as a particular case of his study of spaces of functions defined via mean oscillation on cubes. Extending the results of *Peetre, Campanato, John-Nirenberg,* and *Meyers,* Spanne established relations between generalized Campanato-Morrey-Hölder spaces, defined via a growth function $\varphi(r)$, which satisfies suitable conditions.

Later, *Mizuhara* considered generalized Morrey spaces, taking as a growth function an increasing positive measurable function $\Phi(r)$. Assuming that Φ satisfies the doubling condition

$$\Phi(2r) \le c\Phi(r)$$

with a constant $1 \le c < 2^n$, Mizuhara estimates the maximal operator and some classical singular integrals acting in these new generalized Morrey spaces (see [46]. His ideas have been developed further by *Nakai* (see [51, 53]), considering growth functions $\omega(x,r)$ that depend on the ball of integration:

$$\omega(\mathscr{B}_r(x)) = \omega(x,r):\ \mathbb{R}^n \times \mathbb{R}^+ \to \mathbb{R}^+.$$

Definition 1.4. *A function $f \in L^p_{\rm loc}(\mathbb{R}^n)$, $1 \le p < \infty$, belongs to the generalized Morrey space $L^{p,\omega}(\mathbb{R}^n)$ where ω is a positive measurable function defined on $\mathbb{R}^n \times \mathbb{R}^+$ if*

$$\|f\|_{L^{p,\omega}(\mathbb{R}^n)} = \sup_{\mathscr{B}_r(x)} \left(\frac{1}{\omega(x,r)} \int_{\mathscr{B}_r(x)} |f(y)|^p\,dy \right)^{1/p} < \infty$$

where the supremum is taken over all balls $\mathscr{B}_r(x) \subset \mathbb{R}^n$.

Let $\Omega \subset \mathbb{R}^n$ be a bounded domain. Then the space $L^{p,\omega}(\Omega)$ contains functions $f \in L^p(\Omega)$ for which

$$\|f\|_{L^{p,\omega}(\Omega)} = \sup_{\mathscr{B}_r(x),\, r<d_\Omega} \left(\frac{1}{\omega(x,r)} \int_{\Omega_r(x)} |f(y)|^p\,dy \right)^{1/p} < \infty.$$

If $\omega = r^{\lambda}$, where $\lambda \in (0,n)$, then $L^{p,\omega} \equiv L^p$. However, if $\omega = r^n$, then $L^{p,\omega} = L^{\infty}$.

In the same paper, Nakai investigates the behaviour of the Maximal and Calderón-Zygmund operators acting on functions belonging to $L^{p,\omega}$. His results, together with the techniques developed by Chiarenza, Frasca, and Longo (see [26,27]), can be considered fundamental for the further development of the regular theory for PDEs in generalized Morrey spaces.

Theorem 1.7 ([51]). *Assume that there exists a constant $c > 0$ such that, for any fixed $x_0 \in \mathbb{R}^n$ and for any $r > 0$, the following conditions hold:*

$$c^{-1} \le \frac{\omega(x_0,s)}{\omega(x_0,r)} \le c \quad \forall\, r \le s \le 2r \tag{1.9}$$

$$\int_r^{\infty} \frac{\omega(x_0,s)}{s^{n+1}}\, ds \le c\, \frac{\omega(x_0,r)}{r^n}. \tag{1.10}$$

For $q \ge 1$, let $\mathscr{M}_q f(x) = (\mathscr{M}|f|^q(x))^{1/q}$. Then:

(i) For $1 \le q < p < \infty$, there is a constant $c_{p,q} > 0$ such that

$$\|\mathscr{M}_q f\|_{L^{p,\omega}(\mathbb{R}^n)} \le c_{p,q} \|f\|_{L^{p,\omega}(\mathbb{R}^n)} \qquad \forall\, f \in L^{p,\omega}(\mathbb{R}^n).$$

(ii) For $1 \le p < \infty$, there exists a constant $c_p > 0$ such that, for any $t > 0$ and for any ball $\mathscr{B}_r(x)$, it holds that

$$\frac{|\{y \in \mathscr{B}_r(x) : \mathscr{M}_p f(y) > t\}|}{\omega(x,r)} \le \frac{c_p}{t^p} \|f\|^p_{L^{p,\omega}(\mathbb{R}^n)} \qquad \forall\, f \in L^{p,\omega}(\mathbb{R}^n).$$

(iii) Let $\omega(x_0,r)$ satisfy (1.9), and suppose that there exists a constant $\kappa_3 > 0$ such that

$$\int_r^{\infty} \frac{\omega(x_0,t)}{t^{\delta n+1}} dt \le \kappa_3 \frac{\omega(x_0,r)}{r^{\delta n}}, \tag{1.11}$$

where $0 < \delta < 1$. Then there exists a constant $c = c(p,n,\delta)$ such that

$$\|f\|_{p,\omega} \le c \|f^{\sharp}\|_{p,\omega} \qquad \forall\, f \in L^{p,\omega}(\mathbb{R}^n).$$

The above assertions also hold in $\mathbb{R}^n_+$, and the proof follows the same lines as in Theorem 1.7 (see [30,64]).

Example 1.8. *The set of growth functions is not empty, as demonstrated by the following examples:*

(*i*) *Let w be an A_p weight in the sense of Muckenhoupt. Then the following growth function satisfies conditions* (1.9) *and* (1.10)*:*

$$\omega(x,r)=\Big(\int_{\mathscr{B}_r(x)} w(y)\,dy\Big)^{\alpha},\quad 0<\alpha<1,\quad 1<p<1/\alpha.$$

(*ii*) *The function $f(x)=\chi_{[-1,1]}|x|^{-1/2}$ belongs to $L^{1,\omega}(\mathbb{R})$ with*

$$\omega(x,r)=\int_{\mathscr{B}_r(x)} |y|^{\alpha}\,dy,\quad -1<\alpha\le -\frac{1}{2}$$

where $\mathscr{B}_r$ is any interval of length $2r$ in $\mathbb{R}$.

□

More generalizations of Morrey spaces can be found in [34, 38, 63, 78].

In the papers [35–37], we studied the regularity properties of solutions of PDEs in generalized Morrey spaces, defined in the following way.

Definition 1.5. *Let $\varphi\colon \mathbb{R}^n\times\mathbb{R}^+\to\mathbb{R}^+$ be a measurable function, and let $1\le p<\infty$. The space $M^{p,\varphi}(\mathbb{R}^n)$ consists of all functions $f\in L^p_{\mathrm{loc}}(\mathbb{R}^n)$ such that*

$$\|f\|_{M^{p,\varphi}(\mathbb{R}^n)}=\sup_{x\in\mathbb{R}^n,\,r>0}\frac{1}{\varphi(x,r)}\Big(\frac{1}{r^n}\int_{\mathscr{B}_r(x)}|f(y)|^p\,dy\Big)^{1/p}<\infty.$$

For any bounded domain Ω, we define the space $M^{p,\varphi}(\Omega)$ by considering $f\in L^p(\Omega)$ and $\Omega_r(x)$ instead of $\mathscr{B}_r(x)$ in the norm above.

If $\varphi(x,r)=r^{(\lambda-n)/p}$, then $M^{p,\varphi}\equiv L^{p,\lambda}$, if $\varphi(x,r)=\omega(x,r)^{1/p}r^{-n/p}$ were ω satisfies conditions (1.9) and (1.10), we recover the spaces $L^{p,\omega}$ studied by Nakai [51].

The boundedness of the maximal operator in such spaces can be established under suitable conditions on φ (see [19, 37] and the references therein).

Theorem 1.9. *Assume that there exists a positive constant κ such that for any fixed $x\in\mathbb{R}^n$ and any $r>0$, the following condition holds:*

$$\sup_{r<s<\infty}\frac{\operatorname*{essinf}_{s<\sigma<\infty}\varphi(x,\sigma)\sigma^{n/p}}{s^{n/p}}\le\kappa\,\varphi(x,r). \tag{1.12}$$

Then, for all $f\in M^{p,\varphi}(\mathbb{R}^n)$ with $p>1$, there exists a constant $c(p,\kappa)>0$ such that

$$\|f\|_{M^{p,\varphi}(\mathbb{R}^n)}\le\|\mathscr{M}f\|_{M^{p,\varphi}(\mathbb{R}^n)}\le c\,\|f\|_{M^{p,\varphi}(\mathbb{R}^n)}. \tag{1.13}$$

Let $\phi : \mathbb{R}^+ \to \mathbb{R}^+$ be a decreasing function such that $r^{n/p}\phi(r)$ is almost increasing for $r > 0, p > 1$. Then it verifies (1.12), and (1.13) holds in the space $M^{p,\phi}(\mathbb{R}^n)$.

The *weighted generalized Morrey spaces* $L^{p,\kappa}(w)$ are defined for $1 \leq p < \infty$ and $0 < \kappa < 1$, where w is a positive weight belonging to $L^1_{\mathrm{loc}}(\mathbb{R}^n)$ (see [42]). These spaces consist of all functions $f \in L^{p,w}(\mathbb{R}^n)$ such that the following norm is finite:

$$\|f\|_{L^{p,\kappa}_w(\mathbb{R}^n)} = \sup_{\mathscr{B}_r(x)} \left(\frac{1}{w(\mathscr{B}_r(x))^\kappa} \int_{\mathscr{B}_r(x)} |f(y)|^p w(y)\, dy \right)^{1/p}$$

where the supremum is taken over all balls $\mathscr{B}_r(x)$ in $\mathbb{R}^n$.

Example 1.10. *It is easily seen that if $w = 1$ and $\kappa = \lambda/n$ with $0 < \lambda < n$, we recover the classical Morrey spaces. However, the class of admissible weight functions is far from trivial, as illustrated by the following examples:*

1. *If w satisfies the doubling condition $w(\mathscr{B}_{2r}(x)) \leq c\,w(\mathscr{B}_r(x))$ with some constant $c > 0$, then $L^{p,0}_w = L^p_w$ and $L^{p,1}_w = L^\infty_w$ by the Lebesgue differentiation theorem.*
2. *If $w = |x|^\alpha$, $x \in \mathbb{R}$, for some fixed $\alpha \in (-1/2, 0)$, then the function*

$$f(x) = \chi_{(0,a)}(x)|x|^{-1/2}$$

belongs to $L^{1,(\alpha+1/2)/(\alpha+1)}_w \setminus L^{2(\alpha+1)}_w$ (cf. [42]).

□

The Maximal operator with respect to a weighted measure can be defined as:

$$\mathscr{M}_w f(x) = \sup_{\mathscr{B} \ni x} \frac{1}{w(\mathscr{B})} \int_{\mathscr{B}} |f(y)|\, w(y) dy, \quad w(\mathscr{B}) = \int_{\mathscr{B}} w(y) dy$$

where the supremum is taken over all balls containing x.

The operator $\mathscr{M}_w$ is known to be bounded on $L^{p,\kappa}_w(\mathbb{R}^n)$ when $w \in A_p$ for $1 < p < \infty$ and $0 < \kappa < 1$ (see [42]).

In the limiting case $p = 1$, the following *weak-type maximal inequality* holds:

$$w\big(\{x \in \mathscr{B} : \mathscr{M} f(x) > t\}\big) \leq \frac{c}{t} \|f\|_{L^{1,\kappa}_w} w(\mathscr{B})^\kappa$$

for all $t > 0$, any ball $\mathscr{B}$, and any weight $w \in A_1$.

1.4. Vanishing Morrey-type Spaces

Let SS denote the *Schwartz class* of all rapidly decreasing infinitely differentiable functions on $\mathbb{R}^n$, and let SS' denote the set of all tempered distributions on $\mathbb{R}^n$. Furtheremore, let $\mathscr{D} = C_0^\infty$ denote the space of all compactly supported smooth functions in $\mathbb{R}^n$.

It is well known that the classical Morrey spaces $L^{p,\lambda}$ are non-separable Banach spaces. Additionally, the following continuous embedding holds (cf. [17,61]):

$$SS \hookrightarrow B_{p,p}^{\lambda/p} \hookrightarrow L^{\frac{np}{n-\lambda}} \hookrightarrow L^{p,\lambda} \hookrightarrow L_{\rm loc}^{p,\lambda} \hookrightarrow L_{\omega_\alpha}^{p} \hookrightarrow SS'$$

where $B_{p,p}^{\lambda/p}$ is the classical *Besov space,* and $L_{\omega_\alpha}^{p}$ is a *weighted Lebesgue space* with weight

$$\omega_\alpha(x) = (1+|x|^2)^{\alpha/2}, \qquad \alpha < -n/p.$$

However, neither $\mathscr{D}$, nor SS, nor $L^{np/(n-\lambda)}$ is dense in $L^{p,\lambda}$, as demonstrated in the following example.

Example 1.11 (Zorko, [78]). *Let Ω be a bounded* (A)*-domain. For any fixed $x_0 \in \Omega$, the function*

$$f_{x_0}(x) = |x-x_0|^{-(n-\lambda)/p}$$

cannot be approximated by continuous function.

Indeed, for any $g \in C^0(\Omega)$ and any ball $\mathscr{B}_r(x_0) \Subset \Omega$, we have:

$$\begin{aligned}\int_{\mathscr{B}_r(x_0)} |f_{x_0}(y) - g(y)|\, dy &\geq 2^{-p} \int_{\mathscr{B}_r(x_0)} |f_{x_0}(y)|^p\, dy - \int_{\mathscr{B}_r(x_0)} |g(y)|^p\, dy \\ &\geq c(n) r^\lambda \left(\frac{2^{-p}}{\lambda} - \|g\|_{\infty,\mathscr{B}_r(x_0)}^p r^{n-\lambda} \right).\end{aligned}$$

By choosing r sufficiently small, this gives:

$$\frac{1}{r^\lambda} \int_{\mathscr{B}_r(x_0)} |f_{x_0}(y) - g(y)|^p\, dy \geq c(n,p,\lambda) > 0.$$

□

This limitation can be overcome by introducing the so-called *Vanishing Morrey spaces* (cf. [77,78]). By extending f as zero outside Ω, these spaces can be defined over the whole $\mathbb{R}^n$ as:

$$\bar{\mathscr{L}}^{p,\lambda}(\Omega) = \left\{ f \in L^{p,\lambda}(\Omega) : \quad \|f(\cdot - y) - f(\cdot)\|_{L^{p,\lambda}(\Omega)} \to 0,\ |y| \to 0 \right\}. \tag{1.14}$$

The great advantage of these spaces is that C_0^∞ is dense in $\bar{\mathscr{L}}^{p,\lambda}(\Omega)$. In [78], the author describes the properties of the vanishing subspaces $\bar{\mathscr{L}}_\varphi^p(\Omega)$ of a kind of generalized Morrey spaces $\mathscr{L}_\varphi^p(\Omega)$ consisting of all L^p-functions for which the following norm is finite:

$$\|f\|_{\mathscr{L}_\varphi^p(\Omega)} = \sup_{\mathscr{B}_r(x)} \left(\frac{1}{|\mathscr{B}_r(x)|\varphi(r)^p} \int_{\Omega_r(x)} |f(y)|^p \, dy \right)^{1/p} \tag{1.15}$$

where φ is a nonincreasing positive measurable function, and $r^n\varphi(r)^p$ is nondecreasing.

It is shown that $\bar{\mathscr{L}}_\varphi^p$ is the closure of C_0^∞ with respect to the norm (1.15). Moreover, the *vanishing Morrey space* can be described as the dual of an *"anatomic space"* $H^{p,\varphi}$, consisting of all atomic decomposition over functions a_i, each supported in a ball, i.e.,

$$H^{p,\varphi}(\mathbb{R}^n) = \left\{ f(x) = \sum_{i\geq 0} \lambda_i a_i(x), \quad \lambda_i \in \mathbb{R}, \quad \|a_i\|_{L^p(\mathscr{B}_r)} \leq \frac{1}{|\mathscr{B}_r|^{1/q}\varphi(r)} \right\}.$$

The norm in $H^{p,\varphi}(\mathbb{R}^n)$ is given by:

$$\|f\|_{H^{p,\varphi}(\mathbb{R}^n)} = \inf\left\{ \sum_{i\geq 0} |\lambda_i| : \quad f(x) = \sum_{i\geq 0} \lambda_i a_i(x) \right\} < \infty.$$

1.5. Morrey Spaces over Unbounded Domains

The solvability of the Dirichlet problem for linear uniformly elliptic equations has also been studied on an unbounded domains Ω (cf. [76]) with boundaries satisfying an analogue of the (A)-condition, namely,

$$\sup_{x\in\Omega,\, r\in(0,1]} \frac{|\mathscr{B}_r(x)|}{|\Omega_r(x)|} = A_\Omega < \infty. \tag{A'}$$

It is easy to see that (A′) implies the *external cone condition*, that is

$$|\Omega_r(x)| \geq \frac{1}{A_\Omega} r^n, \qquad \forall x \in \Omega, \quad \forall r \in (0,1] \tag{1.16}$$

and, in the case of bounded domains Ω, this is equivalent to (A).

The regularity of the coefficients of the linear elliptic operator considered by *Transirico, Troisi,* and *Vitolo* in the paper [76] is described using Morrey-type spaces $M^{p,\lambda}(\Omega,t)$, defined as the set of all functions $f \in L^p_{\text{loc}}(\Omega)$ for which

$$\|u\|_{M^{p,\lambda}(\Omega,t)} = \sup_{\substack{r\in(0,t] \\ x\in\Omega}} \left(\frac{1}{r^\lambda} \int_{\Omega_r(x)} |f(y)|^p \, dy \right)^{1/p} < \infty \tag{1.17}$$

for every $t>0$, $p\in[1,\infty)$, and $\lambda\in(0,n)$.

This function class is smaller than $L^p_{\mathrm{loc}}(\Omega)$ but larger than the classical Morrey space $L^{p,\lambda}(\mathbb{R}^n)$. Moreover, the following inclusions hold (cf. [20, 44, 58, 76])

$$M^{p_0,\lambda_0}(\Omega,t)\hookrightarrow M^{p,\lambda}(\Omega,t)$$

if $1\le p\le p_0$ and $(n-\lambda)/p\le(n-\lambda_0)/p_0$.

It is easy to see that for all $t_1,t_2>0$ there exist positive constants c_1,c_2 such that

$$c_1\|f\|_{M^{p,\lambda}(\Omega,t_1)}\le\|f\|_{M^{p,\lambda}(\Omega,t_2)}\le c_2\|f\|_{M^{p,\lambda}(\Omega,t_1)}\qquad\forall\, f\in M^{p,\lambda}(\Omega,t_1).$$

Because the norm is equivalent for different choice of the parameter $t>0$, we can write $M^{p,\lambda}(\Omega)$ instead of $M^{p,\lambda}(\Omega,t)$, and denote $M^p(\Omega)$ for $M^{p,0}(\Omega)$.

We also consider certain subspaces that gather functions with specific *"good behaviour"*, which allow us to study regularity properties of PDE solutions.

We denote by $VM^{p,\lambda}(\Omega)$ the *vanishing* $M^{p,\lambda}(\Omega)$*-space* consisting of functions for which

$$\lim_{t\to 0}\|f\|_{M^{p,\lambda}(\Omega,t)}=0$$

and write $VM^p(\Omega)$ for $VM^{p,0}(\Omega)$. The following example shows that this is a strict subspace of $M^{p,\lambda}(\Omega)$.

Example 1.12 ([76]). *Set* $I_k=[k,k+1/k]$, $k\in\mathbb{N}$, $\Omega=((1,\infty])^n$, *and consider the functions*

$$f:(1,\infty)\to\sum_{k=1}^{\infty}k\chi_{I_k}(x),\qquad g:\Omega\to\prod_{i=1}^{n}f(x_i).$$

Then $g\in M^1(\Omega)\setminus VM^1(\Omega)$. □

Using the embedding properties of Morrey spaces, [20, 58] it can be shown that

$$M^{p_0,\lambda_0}(\Omega)\hookrightarrow VM^{p,\lambda}(\Omega)$$

if $1\le p\le p_0$ and $(n-\lambda)/p<(n-\lambda_0)/p_0$, and in particular $L^\infty(\Omega)\subset VM^{p,\lambda}(\Omega)$.

The closure of $L^\infty(\Omega)$ *with respect to the norm* (1.17) is denoted by $\tilde M^{p,\lambda}(\Omega)$. As usual, we write $\tilde M^p(\Omega)$ instead of $\tilde M^{p,0}(\Omega)$. It is easy to see that

$$L^\infty(\Omega)\subset\tilde M^{p,\lambda}(\Omega)=VM^{p,\lambda}(\Omega)\cap\tilde M^p(\Omega)\subset VM^{p,\lambda}(\Omega).$$

The closure of $C_0^\infty(\Omega)$ *with respect to the norm* (1.17) is denoted by $W_0^{p,\lambda}(\Omega)$. This space can also be characterised using functions in $\tilde{M}^{p,\lambda}(\Omega)$. Specifically, if $\xi_r \in C_0^\infty$ satisfies

$$\xi_r(x) = 1 \text{ if } x \in \bar{\mathscr{B}}_r \text{ and } 0 \text{ for } x \notin \mathscr{B}_{2r},$$

then $f \in M_0^{p,\lambda}(\Omega)$ if f belongs to the closure of L^∞ and satisfies

$$\lim_{r\to\infty} \|(1-\xi_r)f\|_{M^{p,\lambda}(\Omega)} = 0. \tag{1.18}$$

As a consequence,

$$M_0^{p,\lambda}(\Omega) = \tilde{M}^{p,\lambda}(\Omega) \cap M_0^p(\Omega).$$

A natural question is: *what is the relation between* $M^{p,\lambda}(\Omega)$ *and the classical Morrey space?*

Considering the extension f_0 of a function $f \in M^{p,\lambda}(\Omega)$ as zero in $\mathbb{R}^n \setminus \Omega$, we have

$$f_0 \in L^{p,\lambda}(\mathbb{R}^n), \qquad \|f_0\|_{L^{p,\lambda}(\mathbb{R}^n)} \le c\|f\|_{M^{p,\lambda}(\Omega)},$$

where the constant c is independent of f and Ω.

Generalized Morrey-type spaces over unbounded domains were introduced to extend the classical framework to more flexible settings (see [7]). For $1 \le p < \infty$, $t > 0$, and a given growth function $\omega\colon \mathbb{R}^n \times \mathbb{R}^+ \to \mathbb{R}^+$, the spaces $M_\omega^p(\Omega,t)$ consists of all functions $f \in L^p_{\text{loc}}(\Omega)$ such that

$$\|f\|_{M_\omega^p(\Omega,t)} = \sup_{x\in\Omega,\, r\in(0,t]} \left(\frac{1}{\omega(x,r)}\int_{\Omega_r(x)} |f(y)|^p\,dy\right)^{\frac{1}{p}} < \infty. \tag{1.19}$$

To define subspaces containing functions with more refined behaviour, additional conditions are imposed on the growth function ω. More precisely, ω is assumed to satisfy:

- *Doubling condition:* There exists a constant $C_\omega > 0$ such that

$$C_\omega^{-1} \le \frac{\omega(x,s)}{\omega(x,r)} \le C_\omega \qquad \forall\, x \in \mathbb{R}^n, \quad r \le s \le 2r; \tag{W_1}$$

- *Monotonicity condition:*

$$w(x,r) \le w(y,s) \qquad \forall\, x,y \in \mathbb{R}^n, \quad \mathscr{B}_r(x) \subseteq \mathscr{B}_s(y). \tag{W_2}$$

If ω satisfies both (W_1) and (W_2), the norm defined in (1.19) becomes independent of the parameter t, so it suffices to write $M^p_\omega(\Omega)$ instead of $M^p_\omega(\Omega,t)$.

- *Measure condition:* Let χ_Ω denote the characteristic function of Ω, then

$$\|\chi_\Omega\|_{M^p_\omega(\Omega)} = \sup_{x\in\Omega,\, r\in(0,t]} \frac{|\Omega_r(x)|}{\omega(x,r)} = D < \infty. \tag{W_3}$$

Applying Hölder's inequality togather with (W_3), it follows that

$$M^p_\omega(\Omega) \subseteq M^q_\omega(\Omega) \qquad \forall\, 1 \le q \le p < \infty. \tag{1.20}$$

In addition, condition (W_3) ensures that

$$L^\infty(\Omega) \subset M^p_\omega(\Omega),$$

that is clear from the estimate

$$\|f\|_{M^p_\omega(\Omega)} \le \|f\|_{L^\infty(\Omega)} \|\chi_\Omega\|_{M^p_\omega(\Omega)} \le C. \tag{1.21}$$

Definition 1.6. *The vanishing subspace $VM^p_\omega(\Omega)$, for $1 \le p < \infty$ and a growth function ω satisfying (W_1), (W_2), and (W_3), consists of all functions $f \in M^p_\omega(\Omega)$ such that*

$$\lim_{t\to 0} \|f\|_{M^p_\omega(\Omega,t)} = 0.$$

As in (1.20), Hölder's inequality and (W_3) imply that

$$VM^p_\omega(\Omega) \subseteq VM^q_\omega(\Omega), \qquad \forall\, 1 \le q \le p < \infty. \tag{1.22}$$

Now, suppose that $\chi_\Omega \in VM^p_\omega(\Omega)$ satisfies the *vanishing condition:*

$$\lim_{t\to 0} \|\chi_\Omega\|_{M^p_\omega(\Omega,t)} = 0. \tag{V}$$

In this case we have the inclusion

$$L^\infty(\Omega) \subset VM^p_\omega(\Omega),$$

which follows directly from (1.21). Moreover, condition (V) allows for a refinement of (1.22), yielding:

$$M^p_\omega(\Omega) \subset VM^q_\omega(\Omega) \qquad \forall\, 1 \le q < p < \infty. \tag{1.23}$$

Example 1.13. *Let $u \in A_p$, with $p > 1$, in Ω. Define the growth function by*

$$\omega(x,r) = \left(\int_{\mathscr{B}_r(x)} u(y)\,dy \right)^{\alpha} \quad with \quad 0 < \alpha \leq 1/p < 1.$$

This choice of ω satisfies conditions (W$_1$), (W$_2$), and (V). Indeed, by the Lebesgue Differentiation Theorem:

$$\lim_{r\to 0} \frac{1}{|\mathscr{B}_r(x)|} \int_{\mathscr{B}_r(x)} u(y)\,dy = u(x) \qquad for\ a.a.\ x \in \Omega.$$

Therefore,

$$\lim_{t\to 0} \sup_{x\in\Omega,\, r\in(0,t]} \frac{|\Omega_r(x)|}{\left(\int_{\mathscr{B}_r(x)} u(y)\,dy\right)^{\alpha}} \leq c \lim_{t\to 0} \sup_{x\in\Omega,\, r\in(0,t]} \frac{r^{n(1-\alpha)}}{\left(\frac{1}{|\mathscr{B}_r(x)|}\int_{\mathscr{B}_r(x)} u(y)\,dy\right)^{\alpha}} = 0.$$

Furthermore, ω satisfies (W$_3$), since $\omega(x,r) > 0$ for every ball $\mathscr{B}_r(x)$. □

Example 1.14. *Consider the growth function defined by*

$$\omega(x,r) = \phi(r),$$

where $\phi(r)$ is an increasing function such that $\phi(r)/r$ is decreasing, and

$$\lim_{r\to 0} \phi(r) = 0, \qquad \lim_{r\to 0} \frac{\phi(r)}{r} = +\infty.$$

This choice of ω satisfies (W$_1$), (W$_2$), (W$_3$), and (V). □

The space $VM^p_\omega(\Omega)$ is separable, as shown by the following lemma.

Lemma 1.15 ([7]). *Let (W$_1$), (W$_2$), and (W$_3$) hold. If $f \in VM^p_\omega(\Omega)$ with* $\operatorname{supp} f \Subset \Omega$, *then*

$$\lim_{y\to 0} \|f(\cdot - y) - f(\cdot)\|_{M^p_\omega(\Omega)} = 0,$$

$$\lim_{h\to+\infty} \|J_h * f - f\|_{M^p_\omega(\Omega)} = 0,$$

where $\{J_h\}_{h\in\mathbb{N}}$ is a sequence of mollifiers in $\mathbb{R}^n$.

Another subspace is the closure of the L^∞ functions with respect to the norm in (1.19).

Definition 1.7. *Denote by $\widetilde{M}^p_\omega(\Omega)$ the class of functions $f \in M^p_\omega(\Omega)$, $1 < p < \infty$, and ω satisfying (W$_1$), (W$_2$), and (W$_3$), such that*

$$\lim_{h\to+\infty} \left(\sup_{E\in\Sigma(\Omega),\, \|f\chi_E\|_{M^p_\omega(\Omega)} \leq \frac{1}{h}} \|\chi_E\|_{M^p_\omega(\Omega)} \right) = 0.$$

The main feature of these spaces is given by the following lemma.

Lemma 1.16 ([7]). *A function $f \in M^p_\omega(\Omega)$ belongs to $\widetilde{M}^p_\omega(\Omega)$ if and only if f belongs to the closure of $L^\infty(\Omega)$ with respect to the norm in $M^p_\omega(\Omega)$.*

The subspace $VM^p_\omega(\Omega)$ turns out to be larger than $\widetilde{M}^p_\omega(\Omega)$, as seen from the following result [7].

Lemma 1.17. *Assume that ω satisfies (W_1), (W_2), (W_3), and (V). Then*

$$\widetilde{M}^p_\omega(\Omega) \subset VM^p_\omega(\Omega) \qquad \forall\, p \in [1,\infty).$$

We are now able to improve the inclusion (1.23).

Lemma 1.18. *Suppose that ω satisfies (W_1), (W_2), and (W_3). Then*

$$M^p_\omega(\Omega) \subset \widetilde{M}^q_\omega(\Omega), \qquad \forall\, 1 \le q < p < \infty. \tag{1.24}$$

Let us consider the cut-off function $\zeta_h \in C_0^\infty(\mathbb{R}^n)$ defined as

$$\zeta_h(x) = \begin{cases} 1 & x \in \mathscr{B}_h(0) \\ 0 & x \notin \mathscr{B}_{2h}(0). \end{cases}$$

Definition 1.8. *Suppose that ω satisfies (W_1), (W_2), and (W_3). Then a function $f \in M^p_\omega(\Omega)$ belongs to $\overset{\circ}{M}{}^p_\omega(\Omega)$ if and only if*

$$f \in \widetilde{M}^p_\omega(\Omega) \qquad \textit{and} \qquad \lim_{h \to +\infty} \|(1-\zeta_h)\, f\|_{M^p_\omega(\Omega)} = 0. \tag{1.25}$$

We can describe $\overset{\circ}{M}{}^p_\omega(\Omega)$ by means of the following density result.

Lemma 1.19. *Let (A') and (V) hold. A function $g \in M^p_\omega(\Omega)$ belongs to $\overset{\circ}{M}{}^p_\omega(\Omega)$ if and only if g belongs to the closure of $C_0^\infty(\Omega)$ with respect to the norm in $M^p_\omega(\Omega)$.*

The *vanishing Morrey spaces at the origin* $V_0L^{p,\lambda}(\Omega)$ is a proper subspace of $L^{p,\lambda}(\Omega)$ consisting of all Morrey functions such that

$$\lim_{r \to 0} \sup_{x \in \Omega} \frac{1}{r^\lambda} \int_{\Omega_r(x)} |f(y)|^p \, dy = 0$$

This space is equivalent to (1.17) in the case of a bounded domain (see [17, 24]).

Note that, when $\lambda = 0$, we have:

$$\bar{\mathscr{L}}^{p,0}(\Omega) \equiv V_0L^{p,0}(\Omega) \equiv L^p(\Omega),$$

while for $\lambda > 0$, the following strict inclusion holds:

$$\bar{\mathscr{L}}^{p,\lambda}(\mathbb{R}^n) \subset V_0L^{p,\lambda}(\mathbb{R}^n) \subset L^{p,\lambda}(\mathbb{R}^n)$$

and in the case of a bounded domain: $\bar{\mathscr{L}}^{p,\lambda}(\Omega) \equiv V_0L^{p,\lambda}(\Omega)$.

In the case $\Omega \equiv \mathbb{R}^n$, the *vanishing Morrey space at infinity* $V_\infty L^{p,\lambda}(\mathbb{R}^n)$ is defined as the subset of all Morrey functions for which (cf. [16, 17]

$$\lim_{r\to\infty} \sup_{x\in\mathbb{R}^n} \frac{1}{r^\lambda} \int_{\mathscr{B}_r(x)} |f(y)|^p\,dy = 0. \tag{1.26}$$

The following example illustrates some inclusions between the subspaces of $L^{p,\lambda}(\mathbb{R}^n)$.

Example 1.20 ([17]). *For $\alpha, \beta > 0$, let*

$$f_{\alpha,\beta}(x) = \begin{cases} \dfrac{1}{|x|^\alpha}, & |x| \le 1, \\ \dfrac{1}{|x|^\beta}, & |x| > 1. \end{cases}$$

Then:

$$f_{\alpha,\beta} \in V_0L^{p,\lambda}(\mathbb{R}^n) \cap V_\infty L^{p,\lambda}(\mathbb{R}^n) \quad \textit{if} \quad \alpha < \frac{n-\lambda}{p} < \beta,$$

$$f_{\alpha,\beta} \in L^{p,\lambda}(\mathbb{R}^n) \quad \textit{but} \quad f_{\alpha,\beta} \notin V_0L^{p,\lambda}(\mathbb{R}^n) \cup V_\infty L^{p,\lambda}(\mathbb{R}^n) \quad \textit{if} \quad \alpha = \frac{n-\lambda}{p} = \beta.$$

□

1.6. Anisotropic Morrey Spaces

Let $\alpha = (\alpha_1, \ldots, \alpha_n)$ be a real vector with $\alpha_i \ge 1$ and $|\alpha| = \sum_{i=1}^{n} \alpha_i$. The *anisotropic metric* $\rho\colon \mathbb{R}^n \to [0,\infty)$ associated with the vector α is defined as the solution of the equation (cf. [32, 66, 74])

$$F(x,\rho(x)) = \sum_{i=1}^{n} \frac{x_i^2}{\rho^{2a_i}} = 1 \qquad \text{for } x \in \mathbb{R}^n \setminus \{0\} \tag{1.27}$$

and $\rho(0) := 0$. Indeed, for any fixed $x \in \mathbb{R}^n \setminus \{0\}$, the function $F(\cdot,\rho)$ is decreasing, and hence the equation (1.27) is uniquely solvable with respect to ρ.

It is easy to check that $\rho(x-y)$ defines an *anisotropic distance* having a homogeneity of mixed type

$$\rho(t^\alpha x) = \rho(t^{\alpha_1}x_1,\dots,t^{\alpha_n}x_n) = t\rho(x) \qquad \text{for any } t>0. \tag{1.28}$$

In fact,

$$1 = \sum_{i=1}^{n} \frac{(t^{\alpha_i}x_i)^2}{\rho(t^\alpha x)^{2\alpha_i}} = \sum_{i=1}^{n} \frac{x_i^2}{\rho(x)^{2\alpha_i}} \frac{(t\rho(x))^{2\alpha_i}}{\rho(t^\alpha x)^{2\alpha_i}}$$

which implies (1.28). Thus, $\mathbb{R}^n$ endowed with the metric (1.27) becomes a *mixed-homogeneity metric space* $(\mathbb{R}^n,\rho)$.

The *anisotropic balls* with respect to this metric, centered at $x_0 = (x_1^0,\dots,x_n^0) \in \mathbb{R}^n$ and of radius $r>0$, are the *ellipsoids*

$$\mathscr{E}_r(x_0) = \left\{ y \in \mathbb{R}^n : \frac{(y_1-x_1^0)^2}{r^{2\alpha_1}} + \cdots + \frac{(y_n-x_n^0)^2}{r^{2\alpha_n}} < 1 \right\}$$

with a Lebesgue measure $|\mathscr{E}_r(x_0)| = c(n)\, r^{|\alpha|}$. In the case $\alpha_1 = \cdots = \alpha_n = 1$, and hence $|\alpha| = n$, the metric ρ coincides with the Euclidean metric and $\mathscr{E}_r \equiv \mathscr{B}_r$.

Let $\omega \colon \mathbb{R}^n \times \mathbb{R}^+ \to \mathbb{R}^+$ be a positive measurable function. For any ellipsoid $\mathscr{E}_r(x_0)$, we use the notation $\omega(\mathscr{E}) \equiv \omega(\mathscr{E}_r(x_0)) = \omega(x_0,r)$.

A function $f \in L^p_{\text{loc}}(\mathbb{R}^n)$, $p \in [1,\infty)$, belongs to $L^{p,\omega}(\mathbb{R}^n)$ if the following norm is finite

$$\|f\|_{L^{p,\omega}(\mathbb{R}^n)} = \sup_{\mathscr{E}} \left(\frac{1}{\omega(\mathscr{E})} \int_{\mathscr{E}} |f(y)|\,dy \right)^{1/p} \tag{1.29}$$

where the supremum is taken over all anisotropic ellipsoids varying in $(\mathbb{R}^n,\rho)$.

If $\omega(\mathscr{E}) = r^\lambda$ with $\lambda \in (0,|\alpha|)$, we obtain the classical Morrey spaces over $(\mathbb{R}^n,\rho)$. However, there exist growth functions like $\omega = r\ln(r+2)$ for which $L^{p,\omega}$ does not coincide with any Morrey space.

The Hardy-Littlewood maximal operator $\mathscr{M}f$ and the Sharp maximal operator defined for Lebesgue integrable functions over $(\mathbb{R}^n,\rho)$ have the form

$$\mathscr{M}f(x) := \sup_{\mathscr{E}\ni x} \frac{1}{|\mathscr{E}|} \int_{\mathscr{E}} |f(y)|\,dy, \qquad f^\#(x) := \sup_{\mathscr{E}\ni x} \frac{1}{|\mathscr{E}|} \int_{\mathscr{E}} |f(y)-f_{\mathscr{E}}|\,dy \tag{1.30}$$

for almost all $x \in \mathbb{R}^n$.

The next results are extensions of the maximal and sharp inequalities (cf. [25, 51, 66, 73]).

Theorem 1.21 (Maximal Inequality). *Suppose that there exist constants $\kappa_1 > 1$ and $\kappa_2 > 0$ such that for any fixed $x_0 \in \mathbb{R}^n$ and for any $r > 0$ it holds*

$$\begin{aligned} \kappa_1^{-1} \le \frac{\omega(x_0,t)}{\omega(x_0,r)} \le \kappa_1 \qquad \forall\, r \le t \le 2r \\ \int_r^\infty \frac{\omega(x_0,t)}{t^{|\alpha|+1}}\,dt \le \kappa_2 \frac{\omega(x_0,r)}{r^{|\alpha|}}. \end{aligned} \tag{1.31}$$

Let $\mathscr{M}_q f := (\mathscr{M}|f|^q)^{1/q}$. Then for all $1 \le q < p < \infty$, there exists a constant $c(p,q)$ such that

$$\|\mathscr{M}_q f\|_{L^{p,\omega}(\mathbb{R}^n)} \le c(p,q)\|f\|_{L^{p,\omega}(\mathbb{R}^n)} \qquad \forall\, f \in L^{p,\omega}(\mathbb{R}^n). \tag{1.32}$$

Proof. Fix $x_0 \in \mathbb{R}^n$ and $r > 0$, and denote by $\chi_{\mathscr{E}_r(x_0)}(x)$ the characteristic function of $\mathscr{E}_r(x_0)$. Let $2^k\mathscr{E}_r(x_0)$, $k \in \mathbb{N}$, denote the ellipsoid with radius $2^k r$ centered at x_0. Direct computation gives

$$\mathscr{M}\chi_{\mathscr{E}_r(x_0)}(x) \le 1$$

for all $x \in \mathbb{R}^n$.

Moreover, if $x \in 2^{k+1}\mathscr{E}_r(x_0) \setminus 2^k\mathscr{E}_r(x_0)$, the maximal function of $\chi_{\mathscr{E}_r(x_0)}$ can be evaluated by

$$\frac{r^{|\alpha|}}{(2^{k+1}r - r)^{|\alpha|}} \le \mathscr{M}\chi_{\mathscr{E}_r(x_0)}(x) \le \frac{r^{|\alpha|}}{(2^k r - r)^{|\alpha|}}$$

which gives $\mathscr{M}\chi_{\mathscr{E}_r(x_0)}(x) \simeq 2^{-k|\alpha|}$.

The result follows by applying the *Fefferman-Stein Lemma* [31, Lemma 1] (see also [25, 33]). Precisely, for any $1 < p < \infty$ there exists a constant c_p such that

$$\int_{\mathbb{R}^n} (\mathscr{M}f(x))^p \chi_{\mathscr{E}_r(x_0)}(x)\,dx \le c_p \int_{\mathbb{R}^n} |f(x)|^p \mathscr{M}(x)\,dx. \tag{1.33}$$

Then by (1.31) and (1.33), we have

$$\begin{aligned} \int_{\mathscr{E}_r(x_0)} (\mathscr{M}_q f(x))^p\,dx &= \int_{\mathbb{R}^n} (\mathscr{M}|f|^q(x))^{\frac{p}{q}} \chi_{\mathscr{E}_r(x_0)}\,dx \\ &\le c \int_{\mathbb{R}^n} |f(x)|^p \mathscr{M}\chi_{\mathscr{E}_r(x_0)}(x)\,dx \\ &\le c \left\{ \int_{2\mathscr{E}_r(x_0)} |f(x)|^p\,dx + \sum_{i=1}^{\infty} \int_{2^{k+1}\mathscr{E}_r(x_0)\setminus 2^k\mathscr{E}_r(x_0)} |f(x)|^p \mathscr{M}\chi_{\mathscr{E}_r(x_0)}(x)\,dx \right\} \end{aligned}$$

$$\leq c\left\{\omega(x_0,2r)+r^{|\alpha|}\sum_{i=1}^{\infty}\frac{\omega(x_0,2^{k+1}r)}{(2^kr)^{|\alpha|}}\right\}\|f\|^p_{L^{p,\omega}(\mathbb{R}^n)}$$

$$\simeq c\left\{\omega(x_0,2r)+r^{|\alpha|}\sum_{i=1}^{\infty}\int_{2^kr}^{2^{k+1}r}\frac{\omega(x_0,t)}{t^{|\alpha|+1}}\,dt\right\}\|f\|^p_{L^{p,\omega}(\mathbb{R}^n)}$$

$$\leq c\,\omega(x_0,r)\|f\|^p_{L^{p,\omega}(\mathbb{R}^n)}$$

where the constant depends on p,q,κ_1, and κ_2. □

Theorem 1.22 (Sharp Inequality). *Let $\delta\in(0,1)$, and suppose that ω satisfies condition* (1.31) *and*

$$\int_r^{\infty}\frac{\omega(x_0,t)}{t^{|\alpha|\delta+1}}\,dt\leq\kappa_3\,\frac{\omega(x_0,r)}{r^{|\alpha|\delta}} \tag{1.34}$$

for some positive constant κ_3. Then, for any $f\in L^1_{\mathrm{loc}}(\mathbb{R}^n)$ and $p\in(1,\infty)$, there exists a constant c_p independent of f such that

$$\|f\|_{L^{p,\omega}(\mathbb{R}^n)}\leq c_p\,\|f^{\#}\|_{L^{p,\omega}(\mathbb{R}^n)}.$$

Proof. Assume that $\omega\in A_\infty$, and that $\mathscr{M}f\in^{q,\omega}(\mathbb{R}^n)$ for some $1<q<\infty$. Then, for each $p\in[q,\infty)$, the following holds (cf. [33, Theorem IV.2.20]):

$$\int_{\mathbb{R}^n}(\mathscr{M}f(x))^p w(x)\,dx\leq c\int_{\mathbb{R}^n}(f^{\#}(x))^p w(x)\,dx. \tag{1.35}$$

Next, by [33, Theorem IV.2.16] and the known inclusions between A_p-classes, we have

$$(\mathscr{M}\chi_{\mathscr{E}})^{\delta}\in A_1\subset A_\infty$$

for any ellipsoid in $\mathbb{R}^n$ and any $0<\delta<1$. This allows the following estimate:

$$J:=\int_{\mathscr{E}_r(x_0)}|f(x)|^p\,dx=\int_{\mathbb{R}^n}|f(x)|^p\chi_{\mathscr{E}_r(x_0)}(x)\,dx$$

$$\leq\int_{\mathbb{R}^n}|\mathscr{M}f(x)|^p(\mathscr{M}\chi_{\mathscr{E}_r(x_0)}(x))^{\delta}\,dx$$

$$\leq c\int_{\mathbb{R}^n}|f^{\#}(x)|^p(\mathscr{M}\chi_{\mathscr{E}_r(x_0)}(x))^{\delta}\,dx$$

$$\leq c\left\{\int_{2\mathscr{E}_r(x_0)}|f^{\#}(x)|^p\,dx+r^{|\alpha|\delta}\sum_{k=1}^{\infty}\frac{1}{(2^kr)^{|\alpha|\delta}}\int_{2^{k+1}\mathscr{E}_r(x_0)\setminus2^k\mathscr{E}_r(x_0)}|f^{\#}(x)|^p\,dy\right\}$$

$$\leq c\left\{\omega(x_0,2r)+r^{|\alpha|\delta}\sum_{k=1}^{\infty}\frac{\omega(x_0,2^{k+1}r)}{(2^kr)^{|\alpha|\delta}}\right\}\|f^{\#}(y)\|^p_{L^{p,\omega}(\mathbb{R}^n)}$$

$$\leq c r^{|\alpha|\delta} \sum_{k=1}^{\infty} \frac{\omega(x_0, 2^k r)}{(2^k r)^{|\alpha|\delta}} \|f^{\#}\|^p_{L^{p,\omega}(\mathbb{R}^n)}.$$

The growth conditions (1.31) and (1.34) give:

$$\frac{\omega(x_0, 2^k r)}{(2^k r)^{|\alpha|\delta}} \simeq \int_{2^k r}^{2^{k+1} r} \frac{\omega(x_0, t)}{t^{|\alpha|\delta+1}}\, dt.$$

Hence,

$$J \leq c r^{|\alpha|\delta} \int_r^{\infty} \frac{\omega(x_0, t)}{t^{|\alpha|\delta+1}}\, dt \|f^{\#}\|_{L^{p,\omega}(\mathbb{R}^n)} \leq c\, \omega(x_0, r) \|f^{\#}\|^p_{L^{p,\omega}(\mathbb{R}^n)}.$$

This completes the proof. □

2. Campanato Spaces

In a series of works, Campanato studied the properties of a three-parameter family of function spaces, denoted by $\mathscr{L}_k^{p,\lambda}(\Omega)$, defined on a bounded domain $\Omega \subset \mathbb{R}^n$, where Ω is an (A)-domain, and a variety of parameters p, λ, k range over a wide spectrum (see [20–22]). His results provide, on one hand, a new characterization of Hölder spaces, and on the other hand, a unification of Lebesgue, Morrey, and Hölder spaces, as well as the Sobolev spaces built upon them, into a single family of function spaces.

2.1. Definitions and Main Properties

Let Ω be a bounded domain in $\mathbb{R}^n$, and assume that Ω satisfies the (A)-condition.

Definition 2.1. *Let $p \geq 1$, $\lambda \geq 0$, and $k \geq 0$, where p and λ are real numbers and k is a non-negative integer. Denote by $\mathscr{P}_k$ the set of all polynomials of degree at most k. A function $f \in L^p(\Omega)$ belongs to the Campanato space $\mathscr{L}_k^{p,\lambda}(\Omega)$ if the following seminorm is finite:*

$$[f]_{\mathscr{L}_k^{p,\lambda}(\Omega)} = \sup_{x\in\Omega, r>0} \left(\frac{1}{r^{\lambda}} \inf_{P\in\mathscr{P}_k} \int_{\Omega_r(x)} |f(y) - P(y)|^p\, dy \right)^{1/p} < \infty. \quad (1.36)$$

In the particular case $k = 0$, the space $\mathscr{L}^{p,\lambda}(\Omega)$ coincides with the spaces introduced by Meyers in [45], where the infimum in (1.36) is achieved when $P \in \mathscr{P}_0$ is equal to the mean value of f over $\Omega_r(x)$, i.e.,

$$[f]_{\mathscr{L}^{p,\lambda}(\Omega)} = \left(\sup_{x\in\overline{\Omega},\, 0<r\leq d_{\Omega}} \frac{1}{r^{\lambda}} \int_{\Omega_r(x)} |f(y) - f_{\Omega_r(x)}|^p\, dy \right)^{1/p}. \quad (1.37)$$

The expression in (1.36) can be completed into a norm, turning $\mathscr{L}_k^{p,\lambda}(\Omega)$ into a Banach space, by setting

$$\|f\|_{\mathscr{L}_k^{p,\lambda}(\Omega)} = \|f\|_{L^p(\Omega)} + [f]_{\mathscr{L}_k^{p,\lambda}(\Omega)}. \tag{1.38}$$

The completeness of the Campanato spaces can be established similarly to that of Morrey spaces.

It is known that the Hölder space $C^{m,\sigma}(\overline{\Omega})$ is a Banach space endowed with the norm

$$|f|_{m,\sigma;\overline{\Omega}} = \sum_{|\beta|=0}^{m} \max_{x\in\overline{\Omega}} |D^\beta f(x)| + \sum_{|\beta|=m} \sup_{x,y\in\overline{\Omega},x\neq y} \frac{|D^\beta f(x) - D^\beta f(y)|}{|x-y|^\sigma}.$$

The following isomorphism result holds (see [20, 22]).

Theorem 2.1. *Let Ω be a domain of* (A)*-type.*

1. *If $0<\lambda<n$ then*

$$\mathscr{L}^{p,\lambda}(\Omega) \cong L^{p,\lambda}(\Omega)$$

and in particular

$$\mathscr{L}^{p,0}(\Omega) = L^p(\Omega).$$

2. *if $n<\lambda\le n+p$, then*

$$\mathscr{L}^{p,\lambda}(\Omega) \cong C^{0,\sigma}(\overline{\Omega}),$$

with $\sigma = \dfrac{\lambda-n}{p}$ and

$$[f]_{0,\sigma;\overline{\Omega}} \le C(A_\Omega, n)[f]_{\mathscr{L}^{p,\lambda}(\Omega)}$$

where

$$[f]_{0,\sigma;\overline{\Omega}} = \sup_{x,y\in\overline{\Omega},x\neq y} \frac{|f(x)-f(y)|}{|x-y|^\sigma}.$$

3. *In the limiting case $\lambda=n$, the space $\mathscr{L}^{p,n}(\Omega)$ satisfies*

$$L^\infty(\Omega) \subset \mathscr{L}^{p,n}(\Omega) \subset L^p(\Omega), \qquad \forall\, p\ge 1$$

and

$$\mathscr{L}^{1,n}(\Omega) = BMO.$$

4. *If* $n+mp<\lambda<n+(m+1)p$, *then*

$$\mathscr{L}_k^{p,\lambda}(\Omega)\cong C^{m,\sigma}(\overline{\Omega})$$

with

$$m=\left[\frac{\lambda-n}{p}\right], \qquad \sigma=\frac{\lambda-n}{p}-m,$$

where $m=0,1,\ldots,k$.

More references regarding the properties and applications of the *classical Morrey-Campanato spaces* and their generalizations can be found in [2,39,43, 54,57–59].

We also note that an analogous result holds in the case where $\Omega=\mathbb{R}^n$ (see [39]). Peetre summarized various results related to the Campanato-Morrey spaces, including interpolation theorems by Stampacchia, Spanne, and others, along with their application to the estimates of certain integral operators (see [57]).

2.2. Generalized Campanato Spaces

Following the initial generalization of the Morrey spaces, a natural question arises: *is it possible to extend these results to generalized Campanato spaces, and under what conditions on the growth function do the embeddings described in Theorem 2.1 hold?*

Spanne first considered this question in [71, 72] . Later, Nakai introduced generalized Morrey, Campanato, and Hölder spaces on *homogeneous metric spaces,* studying the embeddings between these function classes (see [53]). His work was subsequently extended to various generalized spaces under different conditions on the growth functions, and in different metric space settings (cf. [4,23,28,43]).

Let (X,d_X,μ) denote a *homogeneous metric space* equipped with a distance d_X and a non-negative measure μ. Let $1\leq p<\infty$, and let $\phi: X\times\mathbb{R}^+\to\mathbb{R}^+$ be a growth function. For a ball $\mathscr{B}_r(x)$, we write $\phi(\mathscr{B})\equiv\phi(\mathscr{B}_r(x))=\phi(x,r)$.

Let $f\in L^1_{\text{loc}}(X)$. We define the *generalized Campanato, Morrey, and Hölder spaces,* denoted respectively by $\mathscr{L}^{p,\phi}(X)$, $L^{p,\phi}(X)$, and $\Lambda^{\phi}(X)$, with the following norms:

$$\|f\|_{\mathscr{L}^{p,\phi}(X)} = \sup_{\mathscr{B}} \frac{1}{\phi(\mathscr{B})}\left(\frac{1}{\mu(B)}\int_{\mathscr{B}} |f(x)-f_{\mathscr{B}}|\,d\mu\right)^{1/p},$$

$$\|f\|_{L^{p,\phi}(X)} = \sup_{\mathscr{B}} \frac{1}{\phi(\mathscr{B})}\left(\frac{1}{\mu(B)}\int_{\mathscr{B}} |f(x)^p\,dx\right)^{1/p}, \tag{1.39}$$

$$\|f\|_{\Lambda^{\phi}(X)} = \sup_{x,y\in X,\,x\neq y} \frac{2|f(x)-f(y)|}{\phi(x,d_X(x,y))+\phi(y,d_X(x,y))}$$

These quantities are finite. Let $\mathscr{C}$ denote the space of constant functions. Then, the quantity $\|f\|_{\mathscr{L}^{p,\phi}(X)}$ defines a norm on the quotient space $\mathscr{L}^{p,\phi}(X)/\mathscr{C}$. Moreover, for any fixed ball $\mathscr{B}_0$ in X, the expression $\|f\|_{L^{p,\phi}(X)}+|f_{\mathscr{B}_0}|$ defines a norm on $\mathscr{L}^{p,\phi}(X)$, and both are Banach spaces. If $\mu(X)<\infty$, then the norm $\|f\|_{\mathscr{L}^{p,\phi}(X)}+|f_{\mathscr{B}_0}|$ is equivalent to $\|f\|_{\mathscr{L}^{p,\phi}(X)}+\|f\|_{L^p(X)}$.

If the characteristic function of X belongs to $L^{p,\phi}(X)$, then $L^{p,\phi}(X)/\mathscr{C}$ is a Banach space with the norm $\|f-f_{\mathscr{B}_0}\|_{L^{p,\phi}(X)}$, for any choice of the ball $\mathscr{B}_0$.

The space $\Lambda^{\phi}(X)/\mathscr{C}$ is a Banach space with the given norm. Moreover, for any fixed $x_0\in X$, the quantity $\|f\|_{\Lambda^{\phi}(X)}+|f(x_0)|$ defines a norm on $\Lambda^{\phi}(X)$, making it a Banach space as well.

Nakai provided necessary and sufficient conditions on ϕ ensuring the equivalence of these spaces (see [53]). Assume that ϕ satisfies the following conditions:

$$\begin{aligned}
&\frac{1}{\varkappa_1}\le\frac{\phi(x_0,s)}{\phi(x_0,r)}\le\varkappa_1, && 1/2\le s/r\le 2\\
&\frac{\phi(x_0,r)}{r}\le\varkappa_2\frac{\phi(x_0,s)}{s}, && \forall\, 0<s<r\\
&\int_0^r \mu(\mathscr{B}_t(x_0))^{1/p}\frac{\phi(x_0,t)}{t}\,dt\le\varkappa_3\,\mu(\mathscr{B}_r(x_0))^{1/p}\phi(x_0,r), && r>0\\
&\frac{1}{\varkappa_4}\le\frac{\phi(x_0,r)}{\phi(x_1,r)}\le\varkappa_4, && d_X(x_0,x_1)\le r.
\end{aligned} \tag{1.40}$$

Here, $\varkappa_i>0$, $i=1,2,3,4$, are constants independent of r and s, and hold for all points $x_0,x_1\in X$.

Additionally, if $\mu(X)=\infty$, it is assumed that there exists a constant $K>1$ such that

$$\mathscr{B}_{Kr}(x_0)\setminus\mathscr{B}_r(x_0)\neq\emptyset \qquad \forall\, r>r_0\ge 0. \tag{1.41}$$

Theorem 2.2 ([53]). *Suppose that* $\mu(X) = \infty$ *and that conditions* (1.40) *and* (1.41) *hold. Then the following are equivalent:*

(i) There exists a constant $c > 0$ *such that*

$$\int_r^\infty \frac{\phi(x_0,t)}{t}\,dt \le c\,\phi(x_0,r) \quad \forall\, x_0 \in X,\ r > 0;$$

(ii) $\mathscr{L}^{p,\phi}(X)/\mathscr{C} = L^{p,\phi}(X)$, *and*

$$\|f\|_{\mathscr{L}^{p,\phi}(X)} \sim \|f - \lim_{r\to\infty} f_{\mathscr{B}(x_0,r)}\|_{L^{p,\phi}(X)} \quad \forall\, x_0 \in X.$$

In the special case where $X = \mathbb{R}^n$, with d_X being the Euclidean metric and μ the Lebesgue measure, the implication $(i) \Rightarrow (ii)$ was originaly proved by Mizuhara (see [47, 53]).

Theorem 2.3 ([53]). *Let* $\mu(X) = \infty$ *and assume that conditions* (1.40) *and* (1.41) *hold for some* $r_0 \ge 0$. *Then the following are equivalent:*

(i) *There is a constant* $c > 0$ *such that for all* $x_0, x_1 \in X$ *and* $r > 0$,

$$\int_1^{\max\{2,d_X(x_0,x_1),r\}} \frac{\phi(x_0,t)}{t}\,dt + \int_r^{\max\{2,d_X(x_0,x_1),r\}} \frac{\phi(x_1,t)}{t}\,dt \le c\,\phi(x_0,r);$$

(ii) $\mathscr{L}^{p,\phi}(X) = L^{p,\phi}(X)$ *and*

$$\|f\|_{\mathscr{L}^{p,\phi}} + |f_{\mathscr{B}_0}| \sim \|f\|_{L^{p,\phi}(X)};$$

(iii) $1 \in L^{p,\phi}(X)$, *and*

$$\mathscr{L}^{p,\phi}(X)/\mathscr{C} = L^{p,\phi}(X)/\mathscr{C}$$

with the equivalence of the norms

$$\|f\|_{\mathscr{L}^{p,\phi}(X)} \sim \|f - f_{\mathscr{B}_0}\|_{L^{p,\phi}(X)}$$

for every fixed ball $\mathscr{B}_0$.

If $\mu(X) < \infty$, then there exists a constant $R_0 > 0$ such that the space X can be identified with a ball $\mathscr{B}(x, R_0)$ for all $x \in X$. In this case, the inclusions in Theorem 2.3 hold true after replacing condition (i) with the following:

(i') There exists a constant $c > 0$ such that

$$\int_r^{2R_0} \frac{\phi(x_0,t)}{t}\, dt \le c\,\phi(x_0,r), \qquad x_0 \in X,\ 0 < r \le R_0.$$

For embeddings in generalized Hölder spaces, an additional "monotonicity" assumption is required, namely:

$$\phi(x_0,r) \le \kappa_5 \phi(x_1,s) \qquad \text{whenever} \quad \mathscr{B}(x_0,r) \subset \mathscr{B}(x_1,s). \tag{1.42}$$

Theorem 2.4 ([53]). *If* $1 \le p < \infty$ *and* ϕ *satisfies* (1.40) *and* (1.42)*, then the following are equivalent:*

(i) *There exists a constant* $c > 0$ *such that*

$$\int_0^{d_X(x,y)} \frac{\phi(x,t)}{t}\, dt \le c\,\phi(x, d_X(x,y)) \qquad \forall\, x,y \in X;$$

(ii) $\mathscr{L}^{p,\phi}(X) = \Lambda^{\phi}(X)$ *and*

$$\|f\|_{\mathscr{L}^{p,\phi}} + |f_{\mathscr{B}_0}| \sim \|f\|_{\Lambda^{\phi}(X)} + |f(x_0)|;$$

(iii) $\mathscr{L}^{p,\phi}(X)/\mathscr{C} = \Lambda^{\phi}(X)/\mathscr{C}$ *with the equivalence of norms*

$$\|f\|_{\mathscr{L}^{p,\phi}(X)} \sim \|f\|_{\Lambda^{\phi}(X)}.$$

Recently *Cavaliere, Cianchi, Pick, and Slavikova* (see [23]) studied the embeddings of Sobolev spaces built upon general rearrangement-invariant norms into Morrey and Campanato-type spaces $M^{\varphi}(\Omega)$ and $\mathscr{L}^{\varphi}(\Omega)$, where $\varphi\colon (0,\infty) \to (0,\infty)$ is an *admissible* function, meaning that

$$\inf_{r\in[a,\infty)} \varphi(r) > 0 \qquad \forall\, a \in (0,\infty).$$

The *generalized Morrey* and *Campanato spaces* are defined as collections of measurable functions $u\colon \Omega \to \mathbb{R}$ for which the following quantities are finite:

$$\|u\|_{M^{\varphi}(\Omega)} = \sup_{\mathscr{B}\subset\Omega} \frac{1}{\varphi(|\mathscr{B}|^{1/n})} \frac{1}{|\mathscr{B}|} \int_{\mathscr{B}} |u(x)|\, dx$$

$$[u]_{\mathscr{L}^{\varphi}(\Omega)} = \sup_{\mathscr{B}\subset\Omega} \frac{1}{\varphi(|\mathscr{B}|^{1/n})} \frac{1}{|\mathscr{B}|} \int_{\mathscr{B}} |u(x) - u_{\mathscr{B}}|\, dx,$$

where $u_{\mathscr{B}}$ denotes the mean value of u over the ball $\mathscr{B}$.

Given $m \in \mathbb{N}$, the notation $W^m X(\Omega)$ stands for the Sobolev space of functions whose weak derivatives up to the order m belong to the rearrangement-invariant space $X(\Omega)$ associated with the rearrangement-invariant function norm $\|\cdot\|_{X(0,1)}$. Moreover, $\|\cdot\|_{X'(0,1)}$ denotes the associate function norm of $\|\cdot\|_{X(0,1)}$.

The authors established the following *necessary and sufficient* conditions for the above-mentioned embeddings under minimal regularity assumptions on Ω.

Theorem 2.5 ([23]). *Assume that $\Omega \subset \mathbb{R}^n$ is a bounded John domain and $m \in \mathbb{N}$. Let $\|\cdot\|_{X(0,1)}$ be a rearrangement-invariant function norm, and let φ be an admissible function. Then:*

- *The embedding $W^m X(\Omega) \to M^{\varphi}(\Omega)$ holds if and only if*

$$\sup_{r \in (0,1)} \frac{1}{\varphi(r)} \|s^{-1+\frac{m}{n}} \chi_{(r^n,1)}(s)\|_{X'(0,1)} < \infty;$$

- *The embedding $W^m X(\Omega) \to \mathscr{L}^{\varphi}(\Omega)$ holds if and only if*

$$\sup_{r \in (0,1)} \frac{r}{\varphi(r)} \|s^{-1+\frac{m-1}{n}} \chi_{(r^n,1)}(s)\|_{X'(0,1)} < \infty.$$

The results obtained in the paper [23] are much deeper, providing criteria for the embeddings of Sobolev spaces $W^m X(\Omega)$ into higher-order Campanato-type spaces $\mathscr{L}^{k,\varphi}(\Omega)$, where $k \in \{1, 2, \ldots, m-1\}$. Moreover, the optimal target and domain spaces for the relevant embeddings are also identified.

Chapter 2

On Meir-Keeler Condensing Operators and Solutions of Infinite System of Integral Equations in N-variable in Banach Sequence Spaces c_0 and ℓ_1

M. Mursaleen [1] *
S. Kumar[2] *
M. Simbeye[3] *
M. Mpimbo[4] *
[1]Department of Mathematics, Aligarh Muslim University, India
[2]Department of Mathematics, School of Physical Sciences, North-Eastern Hill University, Shillong, Meghalaya-793022, India
[3]Department of Mathematics, University of Dar es Salaam, Tanzania
[4]Department of Mathematics, University of Dar es Salaam, Tanzania

Abstract

In this chapter, we utilize the Meir-Keeler condensing operators to determine the criteria under which the infinite system of integral equations in N variables has a solution in the Banach sequence spaces c_0 and ℓ_1. These findings expand and generalize existing results in the available literature. Furthermore, illustrative examples demonstrate the implications of the established results.

*Corresponding Author's Email: simbeyemesia03@gmail.com, drsengar2002@gmail.com, mursaleenm@gmail.com, kmpimbo33@gmail.com

Keywords: Hausdorff measure of non-compactness, Meir-Keeler condensing operators, System of integral equations, Sequence spaces

AMS Subject Classification: 47H08, 47G15, 47H10.

1. Introduction and Preliminaries

Functional integral equations have various applications in non-linear analysis. Most problems in natural and applied sciences can be formulated using integral equations. On the other hand, differential equations can be converted into integral equations to obtain their solutions.

The study of solutions for functional equations has been the subject of extensive research for a considerable period. Numerous researchers have explored various methodologies to demonstrate the existence of solutions for functional equations, including functional integral equations, differential equations, and integro-differential equations. Measures of non-compactness are important tools in fixed point theory, allowing us to investigate the existence of solutions to functional equations.

In 1930, Kuratowski [79] introduced the concept of a measure of non - compactness, the function that determines the degree of non compactness of a bounded set. Later, Banaś and Goebel [84] generalized this concept into Banach space. In 1955, Darbo [80] used the Kuratowski measure of non-compactness to prove his famous fixed point theorem which was a generalization of the classical Schauder's fixed point theorem and Banach's contraction principle. Darbo's fixed point theorem and its several generalizations, such as Meir-Keeler's fixed point theorem, have been used by various researchers to study the existence of solutions of integral and differential equations in Banach spaces.

The study of sequence spaces has also attracted the interest of many researchers. Kreyszig [88] has shown some sequence spaces that are also Banach spaces. Banaś and Mursaleen [82] discussed measures of non-compactness in some classical sequence spaces. Recently, several researchers studied the existence of solutions of integral equations in sequence spaces.

The study of sequence spaces has also attracted the interest of many researchers. Kreyszig [88] has shown some sequence spaces that are also Banach spaces. Banaś and Mursaleen [82] discussed measures of non-compactness in some classical sequence spaces. Recently, several researchers studied the existence of solutions of integral equations in sequence spaces.

Malik and Jalal [86] investigated the solvability of an infinite system of integral equations of two variables in the spaces c_0 and ℓ_1. They employed Meir-Keeler condensing operators to examine this problem. Samadi *et al.* [87] used measures of non-compactness and a generalization of Darbo's theorem to prove the existence of solutions of integral equations in two variables in Banach spaces c_0 and ℓ_p. Additionally, in the work by Hazarika *et al.* [85], the solvability of an infinite system of integral equations involving two variables was examined in Banach spaces c_0 and ℓ_1. Ghasemi *et al.* [89] made advancements in the findings of Hazarika *et al.* [85]. They demonstrated the existence of solutions for an infinite system of integral equations in N-variables. Their study focused on utilizing Meir-Keeler condensing operators in the space of tempered sequences, specifically c_0^β and ℓ_1^β. This work extends and generalizes the results of [81, 85] in the classical sequence spaces c_0 and ℓ_1.

Consider $\mathbb{R}_+$ as the interval $[0,\infty)$ and assume that $(M, \|.\|)$ is a real Banach space. When referring to a non-empty subset X of M, we use $\bar{X}$ and ConvX to represent the closure and convex closure of X, respectively. The family Q_M encompasses all non-empty and bounded subsets of M, while its subfamily N_M specifically includes all relatively compact sets. Additionally, we consider $C(\mathbb{R}_+^N, \mathbb{R})$ as the space of continuously differentiable functions defined on $\mathbb{R}_+^N$, where $N \in \mathbb{N}$.

Let ω denote all sequences $y = (y_k)_{k=1}^\infty$, real or complex

$$c_0 = \left\{ y \in \omega : y_k \to 0 (k \to \infty), \|y\|_\infty = \sup_k |y_k| \right\}$$

the space of all null sequences, and

$$\ell_1 = \left\{ y \in \omega : \sum_k |y_k| < \infty, \|y\|_1 = \sum_k |y_k| \right\}$$

the space of all absolutely convergent series.

This chapter utilizes the Hausdorff measure of non-compactness to establish the existence of solutions for a system of integral equations involving N variables.

Definition 1.1. *[82] Let $(\mathscr{X}, \rho)$ be a metric space and $\mathscr{Q}$ be a bounded subset of $\mathscr{X}$. Then the Hausdorff measure of non compactness also known as the χ-measure or ball measure of non-compactness of a set $\mathscr{Q}$, denoted by $\chi(\mathscr{Q})$, is defined as the infimum of the set of all reals $\varepsilon > 0$ such that $\mathscr{Q}$ can be covered by a finite number of balls with radii $< \varepsilon$, that is,*

$$\chi(\mathscr{Q}) = \inf \left\{ \varepsilon > 0 : \mathscr{Q} \subset \bigcup_{i=1}^{n} B(x_i, r_i), x_i \in \mathscr{X}, r_i < \varepsilon (i = 1, \ldots, n) n \in \mathbb{N} \right\}.$$

The Hausdorff measure of non-compactness in c_0 is given by

$$\chi(H) = \lim_{n\to\infty}\left\{\sup_{x\in H}\left(\max_{k\geq n}|x_k|\right)\right\},$$

where $x = (x_j)_{j=1}^{\infty} \in c_0$ and $H \in \mathscr{Q}_{c_0}$.

The following is the axiomatic definition of a measure of non-compactness.

Definition 1.2. *[82] A function $\lambda : Q_M \to \mathbb{R}_+$ is called a measure of non-compactness if it satisfies the following conditions:*

i. the family $\ker\lambda = Y \in Q_M : \lambda(Y) = 0$ is nonempty and $\ker\lambda \subset N_M$,

ii. $Y \subset Z \implies \lambda(Y) \leq \lambda(Z)$,

iii. $\lambda(\bar{Y}) = \lambda(Y)$,

iv. $\lambda(ConvY) = \lambda(Y)$,

v. $\lambda(kY + (1-k)Z) \leq k\lambda(Y) + (1-k)\lambda(Z)$ for $k \in [0,1]$,

vi. if (Y_n) is a sequence of closed sets from Q_M such that $Y_{n+1} \subset Y_n$ for $n = 1,2,3,\ldots$ and $\lim_{n\to\infty}\lambda(Y_n) = 0$, then $\bigcap_{n=1}^{\infty} Y_n \neq \emptyset$,

   If a measure of non-compactness satisfies the following additional conditions, then it is called a regular measure:

vii. $\lambda(Y_1 \cup Y_2) = \max\{\lambda(Y_1), \lambda(Y_2)\}$,

viii. $\lambda(Y_1 + Y_2) \leq \lambda(Y_1) + \lambda(Y_2)$,

ix. $\lambda(kY) = |k|\lambda(Y)$,

x. $\ker\lambda = N_M$.

The family *ker*λ is said to be the *kernel* of measure λ.

The following results obtained by Darbo [80] have been beneficial for proving the existence of solutions of functional equations.

Definition 1.3. *Let M_1 and M_2 be two Banach spaces and let λ_1 and λ_2 be arbitrary measures of non-compactness on M_1 and M_2, respectively. An operator f from M_1 to M_2 is called a (λ_1, λ_2)-condensing operator if it is continuous and $\lambda_2(f(D)) < \lambda_1(D)$ for every set $D \in M_1$ with compact closure.*

Remark 1.1. *If $M_1 = M_2$ and $\lambda_1 = \lambda_2 = \lambda$, then f is called a λ-condensing operator.*

Theorem 1.2. *[80] Let H be a nonempty, closed, bounded and convex subset of a Banach space M and $f : H \to H$ be a continuous mapping such that there exists a constant $k \in [0,1)$ with the property $\lambda_2(f(H)) < k\lambda_1(H)$. Then f has a fixed point in H.*

In 1969, Meir and Keeler [83] proved a very interesting fixed point theorem which is also a generalization of Banach's contraction principle and Schauder's fixed point theorem.

Definition 1.4. *[83] Let (X,d) be a metric space. A mapping T on X is considered a Meir-Keeler contraction if, for any $\varepsilon > 0$, there exists $\delta > 0$ satisfying the condition: if $\varepsilon \le d(x,y) < \varepsilon + \delta$, then $d(Tx,Ty) < \varepsilon$ for all $x,y \in X$.*

Theorem 1.3. *[83] Consider a complete metric space (X,d). If $T : X \to X$ is a Meir-Keeler contraction, then T has a fixed point that is unique.*

Aghajan *et al.* [81], generalized the concept of Meir-Keeler contractions to Banach spaces and came up with the following results:

Definition 1.5. *[81] Consider a nonempty subset C of a Banach space M, and let λ be an arbitrary measure of non-compactness defined on M. An operator $T : C \to C$ is referred to as a Meir-Keeler condensing operator if, for any $\varepsilon > 0$, there exists $\delta > 0$ satisfying the following condition:*

$$\varepsilon \le \lambda(X) < \varepsilon + \delta \implies \lambda(T(X)) < \varepsilon$$

for any bounded subset X of C.

Theorem 1.4. *[81] Consider a nonempty subset C of a Banach space M, and let λ be an arbitrary measure of non-compactness defined on M. If the operator $T : C \to C$ is both continuous and satisfies the properties of being a Meir-Keeler condensing operator, then it can be guaranteed that T has at least one fixed point. Moreover, the collection of all fixed points of T within C forms a compact set.*

2. Hausdorff Measure of Non-compactness in Banach Spaces c_0 and ℓ_1

In this section, the Hausdorff measure of non-compactness in sequence space c_0 and ℓ_1 in N variables is formulated.

Let H be a bounded subset of the Banach space $(c_0, \|.\|_{c_0})$, and the Hausdorff measure of non-compactness be given by

$$\chi(H)=\lim_{n\to\infty}\left\{\sup_{v(\rho_1,...,\rho_N)\in H}\left(\max_{k\geq n}|v_k(\rho_1,...,\rho_N)|\right)\right\},$$

where $v(\rho_1,...,\rho_N)=(v_j(\rho_1,...,\rho_N))_{j=1}^{\infty}\in c_0$ for each $(\rho_1,..,\rho_N)\in\mathbb{R}_+^N$ and $H\in Q_{c_0}$.

The Hausdorff measure of non-compactness χ in the Banach space ℓ_1 is defined as follows:

$$\chi(H)=\lim_{n\to\infty}\left[\sup_{v(\rho_1,...,\rho_N)\in H}\left(\sum_{k=n}^{\infty}|v_k(\rho_1,...,\rho_N)|\right)\right],$$

where $v(\rho_1,...,\rho_N)=(v_j(\rho_1,...,\rho_N))_{j=1}^{\infty}\in\ell_1$ for each $(\rho_1,...,\rho_N)\in\mathbb{R}_+^N$ and $H\in Q_{\ell_1}$.

Let us examine the following system of integral equations in N variables, where N represents the number of variables involved in the system:

$$\begin{aligned} v(\rho_1,...,\rho_N) &= h_n\Bigg(\rho_1,...,\rho_N,\int_0^{\rho_N}\cdots\int_0^{\rho_1}g_N(\rho_1,...,\rho_N,u_1,...,u_N,\\ &\qquad v(u_1,..,u_N))du_1\ldots du_N,v(\rho_1,...,\rho_N)\Bigg), \end{aligned} \tag{2.1}$$

where $v(\rho_1,...,\rho_N)=(v_j(\rho_1,...,\rho_N))_{j=1}^{\infty},(\rho_1,...,\rho_N)\in\mathbb{R}_+^N$, and $v_j\in C(\mathbb{R}_+^N,\mathbb{R})$ for all $j\in\mathbb{N}$.

We will prove the existence of a solution for the infinite system of integral equation (2.1) in the Banach spaces c_0 and ℓ_1.

3. Solvability of an Infinite System of Integral Equations in N Variables in Sequence Space c_0

Assume that

i. $h_n:\mathbb{R}_+^N\times\mathbb{R}\times\mathbb{R}^{\infty}\to\mathbb{R}(n\in\mathbb{N})$ is continuous with

$$H_n=\sup_{k\geq n}\left\{|h_n(\rho_1,...,\rho_N),0,v_j^0(\rho_1,...,\rho_N)|:\rho_1,...,\rho_N\in\mathbb{R}_+<\infty\right\},$$

where $v^0(\rho_1,...,\rho_N)=(v_j^0(\rho_1,...,\rho_N))_{j=1}^{\infty}$ and $v_j^0(\rho_1,...,\rho_N)=0$ for all $j\in\mathbb{N}$, $(\rho_1,...,\rho_N)\in\mathbb{R}_+^N$.

Additionally, there are continuous functions u_n and $m_n : \mathbb{R}_+^N \to \mathbb{R}_+$ (where $n \in \mathbb{N}$) such that

$$|h_n(\rho_1,...,\rho_N),p(\rho_1,...,\rho_N),v(\rho_1,...,\rho_N) - h_n(\rho_1,...,\rho_N),q(\rho_1,...,\rho_N),\bar{v}(\rho_1,...,\rho_N)|$$

$$\begin{aligned} \leq \quad & u_n(\rho_1,...,\rho_N)\max_{j\geq n}|v_i(\rho_1,...,\rho_N) - \bar{v_j}(\rho_1,...,\rho_N)| \\ & + m_n(\rho_1,...,\rho_N)|p(\rho_1,...,\rho_N) - q(\rho_1,...,\rho_N)| \end{aligned}$$

where $p,q : \mathbb{R}_+^N \to \mathbb{R}$, $v_j(\rho_1,...,\rho_N) = (v_j(\rho_1,...,\rho_N))_{j=1}^\infty, \bar{v_j}(\rho_1,...,\rho_N) = (v_j(\rho_1,...,\rho_N))_{j=1}^\infty \in \mathbb{R}^\infty$.

ii. The function $g_n : \mathbb{R}_+^{2N} \times \mathbb{R}^\infty \to \mathbb{R}$ is continuous and there exists a constant G_n such that

$$G_n = \sup_n \left\{ m_n(\rho_1,...,\rho_N) \left| \int_0^{\rho_N} ... \int_0^{\rho_1} g_n(\rho_1,...,\rho_N,u_1,...,u_N,v(u_1,...,u_N))du_1..du_N \right| \right\},$$

where

$$\rho_1,...,\rho_N,u_1,...,u_N \in \mathbb{R}_+, v(u_1,...,u_N) \in \mathbb{R}^\infty.$$

Also,

$$\begin{aligned} \lim_{\rho_1,...,\rho_N \to \infty} \Big| m_n(\rho_1,...,\rho_N) \int_0^{\rho_N} ... \int_0^{\rho_1} & [g_n(\rho_1,...,\rho_N,u_1,...,u_N,v(u_1,...,u_N)) \\ & - g_n(\rho_1,...,\rho_N,u_1,...,u_N,\bar{v}(u_1,...,u_N))]\,du_1...du_N\Big| = 0. \end{aligned}$$

iii. An operator V on $\mathbb{R}_+^N \times c_0 \to c_0$ is defined as follows:

$$(\rho_1,...,\rho_N,v(\rho_1,...,\rho_N)) \to (Vv)(\rho_1,...,\rho_N),$$

where

$$\begin{aligned} (Vv)(\rho_1,...,\rho_N) = \Big(& h_1(\rho_1,...,\rho_N),x_1(v)(\rho_1,...,\rho_N),v(\rho_1,...,\rho_N),h_2(\rho_1,...,\rho_N), \\ & x_2(v)(\rho_1,...,\rho_N),v(\rho_1,...,\rho_N),...\Big), \end{aligned}$$

where

$$x_n(v) = \int_0^{(\rho_N)} ... \int_0^{(\rho_1)} g_n(\rho_1,...,\rho_N,u_1,...,u_N,v(u_1,...,u_N))du_1 \ldots du_N.$$

iv. As $n \to \infty$, $H_n \to 0$ and $G_n \to 0$. Moreover

$$\sup_n H_n = H,$$

$$\sup_n G_n = G,$$

and $\sup_n \{u_n(\rho_1,...,\rho_N) : (\rho_1,...,\rho_N) \in \mathbb{R}_+\} = \mathscr{U} < \infty$ with $0 < \mathscr{U} < 1$.

Theorem 3.1. *Assuming conditions $(i)-(iv)$, there exists a solution $v(\rho_1,...,\rho_N) = v_j((\rho_1,...,\rho_N))_{j=1}^{\infty}$ to the infinite system 2.1. This solution belongs to the sequence space c_0 for all $(\rho_1,...,\rho_N) \in \mathbb{R}_+$, and each component v_j is a continuous function $v_j \in C(\mathbb{R}_+^N, \mathbb{R})$.*

Proof. Using the assumptions in our theorem and Equation (2.1), for all $\rho_1,...,\rho_N \in \mathbb{R}_+$ we have

$$\|v(\rho_1,...,\rho_N)\|_{c_0} = \max_{n\geq 1}\{|v(\rho_1,...,\rho_N)|\}$$

$$= \max_{n\geq 1}\left\{\left|h_n\left(\rho_1,...,\rho_N, \int_0^{\rho_N}...\int_0^{\rho_1} g_n(\rho_1,...,\rho_N,u_1,...,u_N,\right.\right.\right.$$

$$v(u_1,..,u_N))du_1...du_N, v(\rho_1,...,\rho_N))|\}$$

$$\leq \max_{n\geq 1}\left\{|h_n\left(\rho_1,...,\rho_N, \int_0^{\rho_N}...\int_0^{\rho_1} g_n(\rho_1,...,\rho_N,u_1,...,u_N,\right.\right.$$

$$v(u_1,..,u_N))du_1\dots du_N, v(\rho_1,...,\rho_N)) - h_n(\rho_1,...,\rho_N), 0, v^0(\rho_1,...,\rho_N)|\}$$

$$+\max_{n\geq 1}\{h_n(\rho_1,...,\rho_N), 0, v^0(\rho_1,...,\rho_N)\}$$

$$\leq \max_{n\geq 1}\left\{u_n(\rho_1,...,\rho_N \max_{j\geq n}|v_j(\rho_1,...,\rho_N)|) + m_n(\rho_1,...,\rho_N)\right.$$

$$\left.\left|\int_0^{\rho_N}...\int_0^{\rho_1} g_N(\rho_1,...,\rho_N,u_1,...,u_N,v(u_1,..,u_N))du_1\dots du_N, v(\rho_1,...,\rho_N\right|\right\} + H_1$$

$$\leq \mathscr{U}\,\|v(\rho_1,...,\rho_N)\|_{c_0} + G_1 + H_1$$

$$\leq \mathscr{U}\,\|v(\rho_1,...,\rho_N)\|_{c_0} + G + H.$$

That is $(1-\mathscr{U})\mathscr{U}\,\|v(\rho_1,...,\rho_N)\|_{c_0} \leq G+H$, and so,

$$\mathscr{U}\,\|v(\rho_1,...,\rho_N)\|_{c_0} \leq \frac{G+H}{1-\mathscr{U}} = (r_1).$$

Let $\bar{\mathscr{B}} = \bar{\mathscr{B}}(v^0(\rho_1,...,\rho_N), r_1)$ denote a closed ball centered at $(v^0(\rho_1,...,\rho_N)$ with a radius of r_1. Thus, $\bar{\mathscr{B}}$ is a nonempty, bounded, closed, and convex subset of c_0. We can define an operator $V = (V_j)$ on $C(\mathbb{R}_+^N, \bar{\mathscr{B}})$ using the following expression:

$$(Vv)(\rho_1,...,\rho_N) = (V_jv)(\rho_1,...,\rho_N) = (h_j(\rho_1,...,\rho_N), x_j(v)(\rho_1,...,\rho_N), v(\rho_1,...,\rho_N)),$$

where

$$v(\rho_1,...,\rho_N) = (v_j(\rho_1,...,\rho_N)) \in \bar{\mathscr{B}}$$

and $v_j \in C(\mathbb{R}_+{}^N, \mathbb{R})$ for all $j \in \mathbb{N}$.

From assumption (iii), for each $(\rho_1,...,\rho_N) \in {}^N_+$ we have

$$\lim_{j\to\infty}(Vv_j)(\rho_1,...,\rho_N) = \lim_{j\to\infty} h_j(\rho_1,...,\rho_N, x_j(v)(\rho_1,...,\rho_N), v(\rho_1,...,\rho_N)) = 0.$$

Thus $(Vv)(\rho_1,...,\rho_N) \in c_0$. Using the fact that $\left\|(Vv)(\rho_1,...,\rho_N) - v^0(\rho_1,...,\rho_N)\right\|_{c_0} \le \delta$, we derive that V is a self map on $\bar{\mathscr{B}}$. We need to prove that the operator V is continuous on $C(\mathbb{R}_+^N, \bar{\mathscr{B}})$. To do that, let $\varepsilon > 0$ and consider an arbitrary $y_m(\rho_1,...,\rho_N) = (y_j(\rho_1,...,\rho_N))_{j=1}^{\infty}$, $y(\rho_1,...,\rho_N) = (y_j(\rho_1,...,\rho_N))_{j=1}^{\infty} \in \bar{\mathscr{B}}$ such that

$$\|y_m(\rho_1,...,\rho_N) - y(\rho_1,...,\rho_N)\|_{c_0} < \frac{\varepsilon}{2\mathscr{U}}$$

for sufficiently large m. We require that $\|Vy_m(\rho_1,...,\rho_N) - Vy(\rho_1,...,\rho_N)\|_{c_o} \to 0$, for large values of m. Now, we show that $|Vy_m(v) - Vy(\rho_1,...,\rho_N)| \to 0$ as $m \to \infty$.

Considering condition (i), for each $(\rho_1,...,\rho_N) \in \mathbb{R}_+$ we get

$$\begin{aligned}
&|(Vy_m(\rho_1,...,\rho_N)) - (V_ny)(\rho_1,...,\rho_N)| = \\
&|h_n(\rho_1,...,\rho_N), x_n(y_m(\rho_1,...,\rho_N)), y_m(\rho_1,...,\rho_N) \\
&-h_n(\rho_1,...,\rho_N, x_n(y)(\rho_1,...,\rho_N), y(\rho_1,...,\rho_N))| \\
&\le u_n(\rho_1,...,\rho_N)\max_{j\ge n}|y_{m,j}(\rho_1,...,\rho_N) - y_j| \\
&+m_n(\rho_1,...,\rho_N)|x_n(y_m)(\rho_1,...,\rho_N) - x_n(y)(\rho_1,...,\rho_N)| \\
&\le \mathscr{U}\,\|y_m(\rho_1,...,\rho_N) - y(\rho_1,...,\rho_N)\|_{c_0} + m_n(\rho_1,...,\rho_N) \\
&\left|\int_0^{\rho_N} \cdots \int_0^{\rho_1} g_n(\rho_1,...,\rho_N, u_1,...,u_N, y_m(u_1,..,u_N)) \right. \\
&\left. - \int_0^{\rho_N} \cdots \int_0^{\rho_1} g_n(\rho_1,...,\rho_N, u_1,...,u_N, y(u_1,..,u_N))du_1...du_N\right|. \qquad (2.2)
\end{aligned}$$

By using condition (ii), we choose $\mathscr{T} > 0$ such that $\max(\rho_1, ..., \rho_N) > \mathscr{T}$, and we derive that

$$\Bigg| m_n(\rho_1, ..., \rho_N) \int_0^{\rho_N} \cdots \int_0^{\rho_1} [g_n(\rho_1, ..., \rho_N, u_1, ..., u_N, y_m(u_1, .., u_N)) \\ - \int_0^{\rho_N}, ..., \int_0^{(\rho_N)} g_n(\rho_1, ..., \rho_N, u_1, ..., u_N, y(u_1, .., u_N))] du_1 ... du_N \Bigg| < \frac{\varepsilon}{2}.$$

It follows that $|V_n y_m(\rho_1, ..., \rho_N) - V_n y(\rho_1, ..., \rho_N)| < \varepsilon$. For $\rho_1, ..., \rho_N \in [0, \mathscr{T}]$, let

$$\mathscr{A}_1^{\mathscr{T}} = \sup\{\rho_1 : \rho_1 \in [0, \mathscr{T}]\},$$
$$\mathscr{A}_2^{\mathscr{T}} = \sup\{\rho_2 : \rho_2 \in [0, \mathscr{T}]\},$$
$$\vdots$$
$$\mathscr{A}_N^{\mathscr{T}} = \sup\{\rho_N : \rho_N \in [0, \mathscr{T}]\},$$
$$\mathscr{M}_T = \sup\{m_n(\rho_1, ..., \rho_N) : \rho_1, ..., \rho_N \in [0, \mathscr{T}]\},$$

and

$$gy_m, y = \sup\{|g_n(\rho_1, ..., \rho_N, u_1, ..., u_N, y_m(u_1, ..., u_N)) - \\ g_n(\rho_1, ..., \rho_N, u_1, ..., u_N, y(u_1, ..., u_N))| : \\ \rho_1, ..., \rho_N \in [0, \mathscr{T}], u_1 \in [0, \mathscr{A}_N^{\mathscr{T}}], ..., u_N \in [0, \mathscr{A}_1^{\mathscr{T}}]\Big\}.$$

By using (2.2) we find that

$$|V_n y_m(\rho_1, ..., \rho_N) - V_m y(\rho_1, ..., \rho_N)| < \frac{\varepsilon}{2} + M^T g_{y_m, y} \mathscr{A}_N^T ... \mathscr{A}_1^T.$$

Since g_n is a continuous function on the set $[0, \mathscr{T}] \times ... \times [0, \mathscr{T}] \times c_0$, we have $g_{y_m, y} \to 0$ as $\varepsilon \to 0$. It shows that

$$|V_n y_m(\rho_1, ..., \rho_N) - V_m y(\rho_1, ..., \rho_N)| \to 0 \text{ as } \|y_m(\rho_1, ..., \rho_N) - y(\rho_1, ..., \rho_N)\|_{c_0} \to 0$$

for large values of m.

Thus, V is continuous on $\bar{\mathscr{B}} \subset c_0$.

To conclude our proof, we will establish that V is a Meir-Keeler condensing operator on $\bar{\mathscr{B}}$. Let $\mathscr{Q}$ be a bounded subset of $\bar{\mathscr{B}}$, and let $\varepsilon > 0$ be arbitrary. Our goal is to find $\delta > 0$ such that $\varepsilon \leq \chi(\mathscr{Q}) < \varepsilon + \delta$ implies $\chi(V(\mathscr{Q})) < \varepsilon$. From hypotheses (i) and (iii) we observe that

$$\chi(V(\mathscr{Q})) = \lim_{n\to\infty} \left[\sup_{v(\rho_1, ..., \rho_N) \in \mathscr{Q}} \left\{ \max_{k \geq n} |h_k(\rho_1, ..., \rho_N) x_k(v), v(\rho_1, ..., \rho_N) \right] \right\}$$

$$\leq \lim_{n\to\infty}\left[\sup_{v(\rho_1,\ldots,\rho_N)\in\mathscr{Q}}\left\{\max_{k\geq n}|h_k(\rho_1,\ldots,\rho_N,x_k(v),v(\rho_1,\ldots,\rho_N))-h_k(\rho_1,\ldots,\rho_N),0\right.\right.$$

$$\left.\left.,v^0(\rho_1,\ldots,\rho_N)|+\max_{k\geq n}|h_k(\rho_1,\ldots,\rho_N,0,v_0(\rho_1,\ldots,\rho_N))|\right\}\right]$$

$$\leq \lim_{n\to\infty}\left[\sup_{v(\rho_1,\ldots,\rho_N)\in\mathscr{Q}}\left\{\max_{k\geq n}\left(u_k(\rho_1,\ldots,\rho_N)\max|v_j(\rho_1,\ldots,\rho_N)|+m_k(\rho_1,\ldots,\rho_N)\right.\right.\right.$$

$$\left.\left.\left.\left|\int_0^{\rho_N}\cdots\int_0^{\rho_1}g_k(\rho_1,\ldots,\rho_N,u_1,\ldots,u_N,v(u_1,..,u_N))]du_1\ldots du_N\right|+H_N\right)\right\}\right]$$

$$\leq \mathscr{U}\lim_{n\to\infty}\left[\sup_{v(\rho_1,\ldots,\rho_N)\in\mathscr{Q}}\left\{\max_{j\geq n}|v(u_1\ldots u_N)|+G_N+H_N\right\}\right].$$

Using the fact that $G_n\to 0$ and $H_n\to 0$ as $n\to\infty$, we deduce

$$\chi(V(\mathscr{Q}))<\mathscr{U}\chi(\mathscr{Q}). \tag{2.3}$$

Taking $\delta=\dfrac{\varepsilon(1-\mathscr{U})}{\mathscr{U}}$, from 2.3 we get

$$\varepsilon\leq\chi(\mathscr{Q})<\varepsilon+\delta\implies\chi(V(\mathscr{Q}))<\varepsilon.$$

Consequently, an operator V satisfies the properties of a Meir-Keeler condensing operator on the set $\bar{\mathscr{B}}\subset c_0$. Then, according to Theorem 1.4, V has a fixed point in $\bar{\mathscr{B}}$. As a result, the system (2.1) admits a solution in the space c_0. □

Example 3.2. *Consider the following infinite system of integral equations:*

$$z_n=\frac{1}{\rho_1\rho_2\rho_3}+\sum_{i=1}^{\infty}\left(\frac{z_j(\rho_1,\rho_2,\rho_3)}{2i^2}\right)+\frac{1}{(n^3+1)e^{\rho_1\rho_2\rho_3}}$$
$$\int_0^{\rho_1}\int_0^{\rho_2}\int_0^{\rho_3}\frac{\sin(z_n(u_1,u_2,u_3))+\cos(u_3)\sin(\sum_{i=1}^{\infty}z_i(u_1,u_2,u_3))}{2+\sin(\sum_{i=1}^{\infty}z_i(u_1,u_2,u_3))}du_1du_2du_3, \tag{2.4}$$

where $n\in\mathbb{N}$.

If we set $N=3$, equation (2.4) can be considered as a specific instance of equation (2.1).
We have

$$h_n(\rho_1,\rho_2,\rho_3,x_n(z_n)(\rho_1,\rho_2,\rho_3),z(\rho_1,\rho_2,\rho_3))=\frac{1}{\rho_1\rho_2\rho_3}$$

$$+\sum_{i=1}^{\infty}\left(\frac{z_j(\rho_1,\rho_2,\rho_3)}{2i^2}\right)+\frac{1}{(n^3+1)e^{\rho_1\rho_2\rho_3}}g_n(\rho_1,\rho_2,\rho_3,u_1,u_2,u_3,z(\rho_1,\rho_2,\rho_3))$$

$$=\frac{\sin(z_n(u_1,u_2,u_3))+\cos(u_3)\sin(\sum_{i=1}^{\infty}z_i(u_1,u_2,u_3))}{2+\sin\left(\sum_{i=1}^{\infty}z_i(u_1,u_2,u_3)\right)}.$$

If $z(\rho_1,\rho_2,\rho_3)\in c_0$, then $h_n(\rho_1,\rho_2,\rho_3,x_n(z_n)(\rho_1,\rho_2,\rho_3),z(\rho_1,\rho_2,\rho_3))\in c_0$. Now, if $y(\rho_1,\rho_2,\rho_3)=y_i(\rho_1,\rho_2,\rho_3)\in c_0$, then we have

$$|h_n(\rho_1,\rho_2,\rho_3,x_n(z_n)(\rho_1,\rho_2,\rho_3),z(\rho_1,\rho_2,\rho_3))$$

$$-h_n(\rho_1,\rho_2,\rho_3,x_n(y_n)(\rho_1,\rho_2,\rho_3),y(\rho_1,\rho_2,\rho_3))|$$

$$\leq\sum_{i=n}^{\infty}\frac{1}{2n^2}|z_i(\rho_1,\rho_2,\rho_3)-y_i(\rho_1,\rho_2,\rho_3)|$$

$$+\frac{1}{(n^3+1)e^{\rho_1\rho_2\rho_3}}|x_nz(\rho_1,\rho_2,\rho_3)-x_ny(\rho_1,\rho_2,\rho_3)|$$

$$\leq\left(\sum_{i=n}^{\infty}\frac{1}{2i^2}\right)\max|z_i(\rho_1,\rho_2,\rho_3)-y_i(\rho_1,\rho_2,\rho_3)|$$

$$+\frac{1}{(n^3+1)e^{\rho_1\rho_2\rho_3}}|x_n(z(\rho_1,\rho_2,\rho_3))-x_n(y(\rho_1,\rho_2,\rho_3))|$$

$$\leq\frac{\pi^2}{12}\max_{i\geq n}|z_i(\rho_1,\rho_2,\rho_3)-y_i(\rho_1,\rho_2,\rho_3)|$$

$$+\frac{1}{(n^3+1)e^{\rho_1\rho_2\rho_3}}|x_n(z(\rho_1,\rho_2,\rho_3))-x_n(y(\rho_1,\rho_2,\rho_3))|.$$

Notice that $u_n(\rho_1,\rho_2,\rho_3)=\frac{\pi^2}{12}$, $m_n(\rho_1,\rho_2,\rho_3)=\frac{1}{(n^3+1)e^{\rho_1\rho_2\rho_3}}$.

We have $0<\mathscr{U}<1$ and $H_n=\sup_n\left\{\left|\frac{1}{\rho_1\rho_2\rho_3+n^2}\right|,\rho_1,\rho_2,\rho_3\in\mathbb{R}_+\right\}\leq 1$, that is $H_n\to 0$ as $n\to\infty$. Also

$$G_n=\sup_n\left|\int_0^{\rho_1}\int_0^{\rho_2}\int_0^{\rho_3}\frac{\sin(z_n(u_1,u_2,u_3))+\cos(u_3)\sin(\sum_{i=1}^{\infty}z_i(u_1,u_2,u_3))}{2+\sin(\sum_{i=1}^{\infty}z_i(u_1,u_2,u_3))}du_1du_2du_3\right|$$
$$:\rho_1,\rho_2,\rho_3u_1,u_2,u_3\in\mathbb{R}_+.$$

Since

$$\left|\int_0^{\rho_1}\int_0^{\rho_2}\int_0^{\rho_3}\frac{\sin(z_n(u_1,u_2,u_3))+\cos(u_3)\sin(\sum_{i=1}^{\infty}z_i(u_1,u_2,u_3))}{2+\sin\left(\sum_{i=1}^{\infty}z_i(u_1,u_2,u_3)\right)}du_1du_2du_3\right|$$
$$<2\left|\int_0^{\rho_1}\int_0^{\rho_2}\int_0^{\rho_3}du_1du_2du_3\right|=2\rho_1\rho_2\rho_3.$$

Therefore,

$$G_n = \sup\left\{ \frac{2\rho_1\rho_2\rho_3}{(n^3+1)e^{\rho_1\rho_2\rho_3}} : \rho_1,\rho_2,\rho_3,u_1,u_2,u_3 \in \mathbb{R}_+ \right\} = \frac{2}{(n^3+1)e}.$$

As $n \to \infty$, we have $G_n \to 0$. Furthermore, as $\rho_1,\rho_2,\rho_3 \to \infty$ we get

$$\left| \frac{1}{(n^3+1)e^{\rho_1\rho_2\rho_3}} \int_0^{\rho_1}\int_0^{\rho_2}\int_0^{\rho_3} \{g_n(\rho_1,\rho_2,\rho_3,u_1,u_2,u_3,z(\rho_1,\rho_2,\rho_3)) \right.$$
$$\left. -g_n(\rho_1,\rho_2,\rho_3,u_1,u_2,u_3,y(\rho_1,\rho_2,\rho_3))du_1du_2du_3\} \right| \to 0.$$

Our hypotheses $(i)-(iv)$ are satisfied, and by Theorem 3.1 we conclude that the infinite system 2.3 has a solution in the sequence space c_0.

4. Existence of Solution of an Infinite System of Integral Equations in N-Variables in Sequence Space ℓ_1

In this section, we will show the application of the Hausdorff measure of non-compactness, and we will establish the conditions for an infinite system (2.1) to have a solution in the Banach space ℓ_1.

Assume that the following conditions are satisfied:

a. The function $h_n : \mathbb{R}_+^N \times \mathbb{R} \times \mathbb{R}^\infty$ is continuous with the series

$$\sum_{n=1}^{\infty} |h_n(\rho_1,...,\rho_N),0,v^0(\rho_1,...,\rho_N)|$$

converging to zero for all $\rho_1,...,\rho_N \in \mathbb{R}_+$, and $v^0(\rho_1,...,\rho_N) = (v_j(\rho_1,...,\rho_N))_{j=1}^\infty \in \mathbb{R}^\infty$ with $v_j^0(\rho_1,...,\rho_N) = 0$ for all $j \in \mathbb{N}$. Moreover, there exist continuous functions α_n and $\beta_n : \mathbb{R}_+^N \to \mathbb{R}_+ (n \in \mathbb{N})$ such that

$$\begin{aligned}
&|h_n(\rho_1,\dots,\rho_N,p(\rho_1,\dots,\rho_N),v(\rho_1,\dots,\rho_N)) \\
&-h_n(\rho_1,\dots,\rho_N,q(\rho_1,\dots,\rho_N),\bar{v}(\rho_1,\dots,\rho_N))| \\
&\le \alpha_n(\rho_1,\dots,\rho_N)|v_n(\rho_1,\dots,\rho_N)-\bar{v}(\rho_1,\dots,\rho_N)| \\
&+\beta_n(\rho_1,\dots,\rho_N)|p(\rho_1,\dots,\rho_N)-q(\rho_1,\dots,\rho_N)|,
\end{aligned}$$

where $p,q : \mathbb{R}_+^N \to \mathbb{R}$, $v(\rho_1,\dots,\rho_N) = (v_j(\rho_1,\dots,\rho_N))_{j=1}^\infty$, and $\bar{v}(\rho_1,\dots,\rho_N) = (\bar{v}j(\rho_1,\dots,\rho_N))_{j=1}^\infty$ are in $\mathbb{R}^\infty$.

b. $g_n : \mathbb{R}_+^{2N} \times \mathbb{R}^\infty \to \mathbb{R}(n \in \mathbb{N})$ is continuous and there exists a constant H_k such that

$$\begin{aligned} H_k &= \sup\left\{\sum_{n\geq}^{\infty}\left(\beta_n(\rho_1,...,\rho_N)\left|\int_0^{\rho_N}\cdots\int_0^{\rho_1}(g_n(\rho_1,...,\rho_N,u_1,...,u_N,\right.\right.\right.\\ &\quad v(u_1,...,u_N)))du_1...du_N|)\},\end{aligned}$$

where $v(u_1,...,u_N) \in \mathbb{R}^\infty, \rho_1,...,\rho_N, u_1,...,u_N \in \mathbb{R}_+$. Also,

$$\begin{aligned}\lim_{\rho_1,...,\rho_N\to\infty} &= \sum_{n=1}^{\infty}\left|\beta_n(\rho_1,...,\rho_N)\int_0^{\rho_N}\cdots\int_0^{\rho_1}(g_n(\rho_1,...,\rho_N,u_1,...,u_N,v(u_1,...,u_N))\right.\\ &\quad -g_n(\rho_1,...,\rho_N,u_1,...,u_N,\bar{v}(\rho_1,...,\rho_N)))du_1...du_N| = 0.\end{aligned}$$

c. Let an operator $V : \mathbb{R}_+^N \times \ell_1 \to \ell_1$ be defined as follows:

$$(\rho_1,\ldots,\rho_N,v(\rho_1,\ldots,\rho_N)) \to (Vv(\rho_1,...,\rho_N)),$$

where

$$\begin{aligned}(Vv)(\rho_1,\ldots,\rho_N) &= (h_1(\rho_1,\ldots,\rho_N,v_1(\rho_1,\ldots,\rho_N),v(\rho_1,\ldots,\rho_N)),h_2(\rho_1,\ldots,\rho_N,\\ &\quad v_2(\rho_1,\ldots,\rho_N),v(\rho_1,\ldots,\rho_N)),\ldots)\end{aligned}$$

and

$$x_n(v)(\rho_1,...,\rho_N) = \int_0^{\rho_N}\cdots\int_0^{\rho_1} g_n(\rho_1,...,\rho_N,u_1,...,u_N,v(u_1,...,u_N))du_1,...,du_N.$$

d. $\lim_{k\to\infty} H_k = 0,\ \sup_k H_k = H,$

$$\gamma_1 = \sup\left\{\sum_{n=1}^{\infty}\beta_n(\rho_1,...,\rho_N) : \rho_1,...,\rho_N \in \mathbb{R}_+\right\},$$

and $\sup_n \alpha_n(\rho_1,...,\rho_N) : \rho_1,...,\rho_N \in \mathbb{R}_+ = \alpha < \infty$ with $0 < \alpha < 1$.

Theorem 4.1. *Under the assumptions $(a)-(d)$, we can establish that the infinite system (2.1) has a solution $v(\rho_1,...,\rho_N) = (v_j(\rho_1,...,\rho_N))_{j=1}^{\infty}$. This solution belongs to the space ℓ_1 and holds true for all $\rho_1,...,\rho_N \in \mathbb{R}_+$. Additionally, each component v_j of the solution belongs to the space $C(\mathbb{R}_+^N,\mathbb{R})$.*

Proof. Using assumptions $(a)-(d)$ and equation (2.1), we have

$$\|v(t1,...,tN)\|_{\ell_1} = \sum_{n=1}^{\infty}|v_n(\rho_1,...,\rho_N)|$$

$$= \sum_{n=1}^{\infty} \left| h_n(\rho_1,...,\rho_N, \int_0^{\rho_N} ... \int_0^{\rho_1} g_n(\rho_1,...,\rho_N,u_1,...,u_N, \right.$$

$$\left. v(u_1,...,u_N))du_1...du_N, v(\rho_1,...,\rho_N) \right|$$

$$\leq \sum_{n=1}^{\infty} \left| h_n(\rho_1,...,\rho_N, \int_0^{\rho_N} ... \int_0^{\rho_1} g_n(\rho_1,...,\rho_N,u_1,...,u_N, \right.$$

$$\left. v(u_1,...,u_N))du_1...du_N, v(\rho_1,...,\rho_N) - h_n(\rho_1,...,\rho_N,0,v^0(\rho_1,...,\rho_N)) \right|$$

$$+ \sum_{n=1}^{\infty} |h_n(\rho_1,...,\rho_N,0,v^0(\rho_1,...,\rho_N))|$$

$$\leq \sum_{n=1}^{\infty} \alpha_n(\rho_1,...,\rho_N)|v_N(\rho_1,...,\rho_N)| + \sum_{n=1}^{\infty} \beta_n(\rho_1,...,\rho_N)$$

$$\left| \int_0^{\rho_N} ... \int_0^{\rho_1} g_n(\rho_1,...,\rho_N,u_1,...,u_N,v(\rho_1,...,\rho_N))du_1...du_N \right|$$

$$\leq \alpha \sum_{n=1}^{\infty} |vn(\rho_1,...,\rho_N)| + H_1 \leq \alpha \|v(\rho_1,...,\rho_N)\| \ell_1 + H.$$

That is $(1-\alpha)||v(\rho_1,...,\rho_N||_{\ell_1} \leq H \implies ||v(t_1,...,t_N||_{\ell_1} \leq \dfrac{H}{1-\alpha} = r_0(say).$

Let us consider $\bar{\mathscr{D}} = \bar{\mathscr{D}}(v^0(\rho_1,...,\rho_N),r_0)$, which denotes a closed ball centered at $v^0(\rho_1,...,\rho_N)$, with a radius of r_0 in the space ℓ_1. As a result, $\bar{\mathscr{D}}$ can be described as a nonempty, bounded, closed, and convex subset of ℓ_1.

Let us define an operator $V = (V_j)$ on $C(\mathbb{R}_+^N, \bar{\mathscr{D}})$ by the formula

$$(Vv)(\rho_1,...,\rho_N) = (V_j v)(\rho_1,...,\rho_N)_{j=1}^{\infty}$$

$$= (h_i(\rho_1,...,\rho_N,x_i(v)(\rho_1,...,\rho_N),v(\rho_1,...,\rho_N)))_{i=1}^{\infty},$$

where $v(\rho_1,...,\rho_N) = (v_j(\rho_1,...,\rho_N))_{j=1}^{\infty} \in \bar{\mathscr{D}}$ and $v_j \in C(\mathbb{R}_+^N,\mathbb{R})$, $\forall j \in \mathbb{N}$.

Using condition (c), for each $(\rho_1,...,\rho_N) \in \mathbb{R}_+^N$ we have

$$\sum_{n=1}^{\infty} \left|(V_j V)(\rho_1,...,\rho_N)\right| = \sum_{n=1}^{\infty} |(h_i(\rho_1,...,\rho_N,x_i(v)(\rho_1,...,\rho_N),v(\rho_1,...,\rho_N)))| < \infty,$$

so $(Vv)(\rho_1,...,\rho_N) \in \ell_1$.

Also, $||V(v)(\rho_1,...,\rho_N) - v^0(\rho_1,...,\rho_N)||_{\ell_1} < r_0$ and so the operator V is a self mapping on $\bar{\mathscr{D}}$.

Now, we are supposed to show that V is continuous. Take

$$y_m(\rho_1,...,\rho_N) = (y_{m,j}(\rho_1,...,\rho_N))_{j=1}^{\infty}, y(\rho_1,...,\rho_N) = (y(\rho_1,...,\rho_N))_{j=1}^{\infty} \in \ell_1$$

and let $\varepsilon > 0$ be arbitrary with $\|y_m(\rho_1,...,\rho_N) - y(\rho_1,...,\rho_N)\|_{\ell_1} < \frac{\varepsilon}{2}$ for large values of m.

We argue that $\|Vy_m(\rho_1,...,\rho_N) - Vy(\rho_1,...,\rho_N)\|_{\ell_1} \to 0$ as $m \to \infty$.

To prove this claim, we will show that $|V_n y_m(\rho_1,...,\rho_N) - V_n y(\rho_1,...,\rho_N)| \to 0$ as $m \to \infty$.

Then for each $(\rho_1,...,\rho_N) \in \mathbb{R}_+$ we have

$$\begin{aligned}
&|(V_n y_m)(\rho_1,...,\rho_N) - (V_n y)(\rho_1,...,\rho_N)| \\
&= |h_n(\rho_1,...,\rho_N, x_n(y_m)(\rho_1,...,\rho_N), y_m(\rho_1,...,\rho_N)) \\
&\quad - h_n(\rho_1,...,\rho_N, x_n(y)(\rho_1,...,\rho_N), y(\rho_1,...,\rho_N))| \\
&\leq \alpha_n(\rho_1,...,\rho_N)|y_m(\rho_1,...,\rho_N) - y(\rho_1,...,\rho_N)| \\
&\quad + \beta_n(\rho_1,...,\rho_N)|x_n(y_m)(\rho_1,...,\rho_N) - x_n(y)(\rho_1,...,\rho_N)| \\
&\leq \alpha|y_m(\rho_1,...,\rho_N) - y(\rho_1,...,\rho_N)| + \beta_n(\rho_1,...,\rho_N) \\
&\quad \left|\int_0^{\rho_N} \cdots \int_0^{\rho_1} [g_n(\rho_1,...,\rho_N, u_1,...,u_N, y_m(u_1,...,u_N)) \right. \\
&\quad \left. - g_n(\rho_1,...,\rho_N, u_1,...,u_N, y(u_1,...,u_N))]\, du_1...du_N \right|.
\end{aligned}$$

And so,

$$\begin{aligned}
&\sum_{n=1}^{\infty} |V_n y_m(\rho_1,...,\rho_N) - V_n y(\rho_1,...,\rho_N)| \\
&\leq \alpha \|y_m(\rho_1,...,\rho_N) - y(\rho_1,...,\rho_N)\|_{\ell_1} + \sum_{n=1}^{\infty} \beta_n(\rho_1,...,\rho_N) \\
&\quad \left|\int_0^{\rho_N} \cdots \int_0^{\rho_1} [g_n(\rho_1,...,\rho_N, u_1,...,u_N, y_m(u_1,...,u_N)) \right. \\
&\quad \left. - g_n(\rho_1,...,\rho_N, u_1,...,u_N, y(u_1,...,u_N))]\, du_1...du_N \right|.
\end{aligned}$$

In view of assumption (b), there exists $\mathscr{T} > 0$ such that if $\max(\rho_1,...,\rho_N) > \mathscr{T}$, then

$$\begin{aligned}
&\sum_{n=1}^{\infty} \beta_n(\rho_1,...,\rho_N) \left|\int_0^{\rho_N} \cdots \int_0^{\rho_1} [g_n(\rho_1,...,\rho_N, u_1,...,u_N, x_m(u_1,...,u_N)) \right. \\
&\quad \left. - g_n(\rho_1,...,\rho_N, u_1,...,u_N, x(u_1,...,u_N))]\, du_1...du_N \right| < \frac{\varepsilon}{2}.
\end{aligned}$$

Hence for $\max(\rho_1,...,\rho_N) > \mathscr{T}$, $\sum_{n=1}^{\infty} |(V_n y_m)(\rho_1,...,\rho_N) - (V_n y)(\rho_1,...,\rho_N)| < \varepsilon$ i.e

$$\|(Vy_m)(\rho_1,...,\rho_N) - (Vy)(\rho_1,...,\rho_N)\| < \varepsilon.$$

Let

$$\mathscr{A}_1^{\mathscr{T}} = \sup\{\rho_1 : \rho_1 \in [0,\mathscr{T}]\},$$
$$\mathscr{A}_2^{\mathscr{T}} = \sup\{\rho_2 : \rho_2 \in [0,\mathscr{T}]\},$$
$$\vdots$$
$$\mathscr{A}_N^{\mathscr{T}} = \sup\{\rho_N : \rho_N \in [0,\mathscr{T}]\},$$

and

$$g_{y_m,y} = \sup\{|g_n(\rho_1,...,\rho_N,u_1,...,u_N,y_m(u_1,...,u_N)) - g_n(\rho_1,...,\rho_N,u_1,...,u_N,y(u_1,...,u_N))| : \rho_1,...,\rho_N \in [0,\mathscr{T}], u_1 \in [0,\mathscr{A}_N^{\mathscr{T}}],...,u_N \in [0,\mathscr{A}_1^{\mathscr{T}}]\}.$$

Then

$$\sum_{n=1}^{\infty} |(V_n y_m)(\rho_1,...,\rho_N) - (V_n y)(\rho_1,...,\rho_N)| < \frac{\varepsilon}{2} + g_{y_m,y}\mathscr{A}_N^{\mathscr{T}} ... \mathscr{A}_1^{\mathscr{T}} \gamma_1.$$

From the fact that g_n is continuous on the set $[0,\mathscr{T}]^N \times [0,\mathscr{A}_N^{\mathscr{T}}] \times ... \times [0,\mathscr{A}_1^{\mathscr{T}}] \times \ell_1$, we derive that $g_{y_m}, y \to 0$ as $m \to \infty$. Thus

$$\sum_{n=1}^{\infty} |(V_n y_m)(\rho_1,...,\rho_N) - (V_n y)(\rho_1,...,\rho_N)| \to 0$$

as $\|y_m(\rho_1,...,\rho_N) - y(\rho_1,...,\rho_N)\|_{\ell_1} \to 0$. We conclude that V is a continuous operator on $\bar{\mathscr{D}} \subset \ell_1$.

To establish that V is a Meir-Keeler condensing operator, we proceed as follows. Consider a given value of $\varepsilon > 0$. Our goal is to find a corresponding value of $\delta > 0$ such that whenever we have $\varepsilon < \chi(X) < \varepsilon + \delta$, it implies $\chi(VX) < \varepsilon$, where X is a nonempty bounded subset of $\mathscr{D}$.

Using condition (a) and (c), we deduce

$$\begin{aligned}
\chi(V(X)) &= \lim_{n\to\infty}\left(\sup_{v(\rho_1,...,\rho_N)\in X}\left\{\sum_{k\geq n}|h_k(\rho_1,...,\rho_N,x_k(v)(\rho_1,...,\rho_N),v(\rho_1,...,\rho_N))|\right\}\right)\\
&\leq \lim_{n\to\infty}\left(\sup_{v(\rho_1,...,\rho_N)\in X}\left\{\sum_{k\geq n}|h_k(\rho_1,...,\rho_N,x_k(v)(\rho_1,...,\rho_N),v(\rho_1,...,\rho_N))-\right.\right.\\
&\qquad \left.\left. h_k(\rho_1,...,\rho_N,0,v^0(\rho_1,...,\rho_N))| + \sum_{k\geq n}|h_k(\rho_1,...,\rho_N,0,v^0(\rho_1,...,\rho_N))|\right\}\right)\\
&\leq \lim_{n\to\infty}\left(\sup_{v(\rho_1,...,\rho_N)\in X}\left\{\sum_{k\geq n}\alpha_k(\rho_1,...,\rho_N)|v(\rho_1,...,\rho_N)| + \beta_k(\rho_1,...,\rho_N)\right.\right.\\
&\qquad \left.\left.\left|\int_0^{\rho_N}...\int_0^{\rho_1} g_k(\rho_1,...,\rho_N,u_1,...,u_N,v(u_1,...,u_N))du_1...du_N\right|\right\}\right)\\
&\leq \lim_{n\to\infty}\left(\sup_{v(\rho_1,...,\rho_N)\in X}\left\{\alpha\sum_{k\geq n}|v(\rho_1,...,\rho_N)| + H_n\right\}\right).
\end{aligned}$$

Using the fact that $H_n \to 0$ as $n \to \infty$, we derive that

$$\chi(V(X)) \leq \alpha\chi(X). \tag{2.5}$$

By choosing $\delta = \dfrac{\varepsilon(1-\alpha)}{\alpha}$, from equation (2.5) it becomes evident that the operator V defined on $\bar{\mathscr{D}} \in \ell_1$ is a Meir-Keeler condensing operator. Hence, based on the statement given in Theorem 1.4, we can conclude that there exists a fixed point for V in $\bar{\mathscr{D}}$. Consequently, we can infer that the infinite system (2.1) has at least one solution within the Banach space ℓ_1. □

Example 4.2. *Consider the following infinite system of integral equations;*

$$z_n(\rho_1,\rho_2,\rho_3) = \sum_{j=n}^{\infty} \frac{\sin(\rho_1\rho_2\rho_3) z_j(\rho_1,\rho_2,\rho_3)}{3j^2}$$
$$+ \frac{1}{n^3 e^{\rho_1\rho_2\rho_3}} \int_0^{\rho_1}\int_0^{\rho_2}\int_0^{\rho_3} \frac{\cos\left(\sum_{j=1}^{\infty} z_j(u_1,u_2,u_3)\right)}{2+\sin(z_n(u_1,u_2,u_3))} du_1 du_2 du_3, \tag{2.6}$$

where $n \in \mathbb{N}$. Here we have

$$h_n(\rho_1,\rho_2,\rho_3,x_n(z(\rho_1,\rho_2,\rho_3)),z(\rho_1,\rho_2,\rho_3)) = \sum_{j=n}^{\infty} \frac{\sin(\rho_1\rho_2\rho_3) z_j(\rho_1,\rho_2,\rho_3)}{3j^2}$$
$$+ \frac{1}{n^3 e^{\rho_1\rho_2\rho_3}} x_n(z_n(\rho_1,\rho_2,\rho_3),$$

where

$$x_n(z(\rho_1,\rho_2,\rho_3)) = \int_0^{\rho_1}\int_0^{\rho_2}\int_0^{\rho_3} g_n(\rho_1,\rho_2,\rho_3,u_1,u_2,u_3,z(\rho_1,\rho_2,\rho_3)),$$

and

$$g_n(\rho_1,\rho_2,\rho_3,u_1,u_2,u_3,z(\rho_1,\rho_2,\rho_3)) = \frac{\cos\left(\sum_{j=1}^{\infty} z_j(u_1,u_2,u_3)\right)}{2+\sin(z_n(u_1,u_2,u_3))}.$$

If $z(\rho_1,\rho_2,\rho_3) \in \ell_1$, then

$$\begin{aligned}
&\sum_{n=1}^{\infty} |h_n(\rho_1,\rho_2,\rho_3,u_1,u_2,u_3,x_n(z(,\rho_2,\rho_3),z(\rho_1,\rho_2,\rho_3)))| \\
\leq\ & \sum_{n=1}^{\infty}\sum_{j=1}^{\infty} \left|\frac{\sin(\rho_1\rho_2\rho_3) z_j(\rho_1,\rho_2,\rho_3)}{3j^2}\right| + \frac{\rho_1\rho_2\rho_3}{e^{\rho_1\rho_2\rho_3}} + \sum_{n=1}^{\infty}\frac{1}{n^3} \\
\leq\ & \sum_{n=1}^{\infty}\sum_{j=n}^{\infty} \left|\frac{z_j(\rho_1,\rho_2,\rho_3)}{3j^2}\right| + \frac{1}{e}\sum_{n=1}^{\infty}\frac{1}{n^3} \\
\leq\ & \|z(\rho_1,\rho_2,\rho_3)\|_{\ell_1} + \frac{\pi}{6e} < \infty
\end{aligned}$$

Therefore, $(h_n(\rho_1,\rho_2,\rho_3,x_n(z_n(\rho_1,\rho_2,\rho_3)),z(\rho_1,\rho_2,\rho_3))) \in \ell_1$.
Now if $y(\rho_1,\rho_2,\rho_3) = (y_j(\rho_1,\rho_2,\rho_3)) \in \ell_1$, then

$$\begin{aligned}
&|h_n(\rho_1,\rho_2,\rho_3,x_n(z(\rho_1,\rho_2,\rho_3)),z(v)) - h_n(\rho_1,\rho_2,\rho_3,x_n(y(\rho_1,\rho_2,\rho_3)),y(\rho_1,\rho_2,\rho_3)))| \\
\leq\ & \sum_{j=n}^{\infty} \frac{1}{3j^2}|z_j(\rho_1,\rho_2,\rho_3) - y_j(\rho_1,\rho_2,\rho_3)| + \frac{1}{n^3 e^{\rho_1\rho_2\rho_3}}|x_n(z(\rho_1,\rho_2,\rho_3)) - x_n(y(\rho_1,\rho_2,\rho_3))| \\
\leq\ & \frac{\pi^2}{12}|x_n(\rho_1,\rho_2,\rho_3) - y_n(\rho_1,\rho_2,\rho_3)| + \frac{1}{n^3 e^{\rho_1\rho_2\rho_3}}|x_n(z(\rho_1,\rho_2,\rho_3)) - x_n(y(\rho_1,\rho_2,\rho_3))|
\end{aligned}$$

We have $\alpha_n(\rho_1,\rho_2,\rho_3) = \dfrac{\pi^2}{12}$ and $\beta_n(\rho_1,\rho_2,\rho_3) = \dfrac{1}{n^3 e^{\rho_1\rho_2\rho_3}}$.

It is obvious that $0 < \alpha < 1$ and $\sum_{n\geq 1} \left|h_n(\rho_1,\rho_2,\rho_3,0,v^0(\rho_1,\rho_2,\rho_3))\right|$ converges to zero $\forall \rho_1,\rho_2,\rho_3 \in \mathbb{R}_+$. Again we have

$$\sum_{n\geq k} \beta_n(\rho_1,\rho_2,\rho_3)|x_n(z(\rho_1,\rho_2,\rho_3))| \leq \frac{\rho_1\rho_2\rho_3}{e^{\rho_1\rho_2\rho_3}} \sum_{n\geq k} \frac{1}{n^3}$$

and

$$\mathcal{Q}_k \leq \sup\left\{ \frac{\rho_1\rho_2\rho_3}{e^{\rho_1\rho_2\rho_3}} \sum_{n\geq k} \frac{1}{n^3} : \rho_1,\rho_2,\rho_3,u_1,u_2,u_3 \in \mathbb{R}_+ \right\}.$$

As $k \to \infty$ we get $\sum_{n\geq k} \dfrac{1}{n^3} \to 0$. Thus, $\mathcal{Q}_k \to 0$ as $k \to \infty$ and $\mathcal{Q} = \dfrac{\pi^2}{6e}$.

Now,

$$\begin{aligned}
&\sum_n \left| \beta_n(\rho_1,\rho_2,\rho_3) \int_0^{\rho_1}\int_0^{\rho_2}\int_0^{\rho_3} [g_n(\rho_1,\rho_2,\rho_3,u_1,u_2,u_3,v(u_1,u_2,u_3)) \right. \\
&\left. -g_n(\rho_1,\rho_2,\rho_3,u_1,u_2,u_3,\bar{v}(\rho_1,\rho_2,\rho_3))]\,du_1du_2du_3 \right| \\
\leq\ & \sum_n \left(\frac{2\rho_1\rho_2\rho_3}{n^3 e^{\rho_1\rho_2\rho_3}} \right) = \frac{\rho_1\rho_2\rho_3}{3e^{\rho_1\rho_2\rho_3}}.
\end{aligned}$$

As $\rho_1,\rho_2,\rho_3 \to \infty$ we have $\dfrac{\rho_1\rho_2\rho_3}{e^{\rho_1\rho_2\rho_3}} \to 0$.

Therefore

$$\begin{aligned}
\lim_{\rho_1,\rho_2,\rho_3\to\infty} &\sum_n \left| \beta_n(\rho_1,\rho_2,\rho_3) \int_0^{\rho_1}\int_0^{\rho_2}\int_0^{\rho_3} [g_n(\rho_1,\rho_2,\rho_3,u_1,u_2,u_3,v(u_1,u_2,u_3)) \right. \\
&\left. -g_n(\rho_1,\rho_2,\rho_3,u_1,u_2,u_3,\bar{v}(\rho_1,\rho_2,\rho_3))]\,du_1du_2du_3 \right| = 0.
\end{aligned}$$

So, all the criteria stated in Theorem 4.1 are fulfilled. Consequently, we can infer that there exists at least one solution within the Banach space ℓ_1 for the infinite system of integral equations (2.6).

5. Conclusion

In this chapter, the conditions for an infinite system of integral equations in N variables have been discussed. To accomplish our objective, we have used Meir-Keeler condensing operators and measures of non-compactness. Our results are the generalizations of several works in the literature on the Banach sequence spaces c_0 and ℓ_1.

Chapter 3

Some Problems of Approximation Theory in Variable Exponent Spaces

Daniyal M. Israfilov[1,2] *
Ahmet Testici[1]
[1]Balikesir University Department of Mathematics,
Balikesir, Turkiye
[2]Institute of Mathematics and Mechanics, The Ministry of Science and Education of Azerbaijan, Baku, Azerbaijan

Abstract

Some problems of approximation theory in variable exponent Lebesgue spaces, defined on intervals of the real line and in Smirnov classes of analytic functions defined on domains of the complex plane are investigated. Moreover, in variable exponent Lebesgue spaces some results related to the approximation properties of matrix transforms are given.

Keywords: Approximation by polynomials, modulus of smoothness, variable exponent Lebesgue spaces, direct and inverse theorems, generalized Lipschitz classes, matrix transforms

AMS Subject Classification: 41A10, 41A30, 42A10, 30E10.

*Corresponding Author's Email: mdaniyal@balikesir.edu.tr

1. Introduction

The aim of this work is to investigate fundamental problems of approximation theory in variable exponent Lebesgue spaces of functions defined on intervals of the real line and various subsets of the complex plane. Additionally, the approximation properties of different summation methods are also investigated.

The important special functions or function classes used in mathematics and its application areas may have quite complicated expressions. In such cases, it is beneficial to approximate these special functions or function classes by simpler functions. Positive solutions to these types of problems facilitate finding approximate expressions for these special functions. Researchers typically consider the space of continuous functions, the Lebesgue integrable function space L_p with a constant p, the Sobolev space W_p and various generalizations of these spaces, as they allow accurate mathematical modeling of certain processes. In contemporary applications in physics and mechanics, some functions may not belong to the classical spaces L_p and W_p. As a result, approximating these functions or finding their approximate expressions can be challenging and problematic, which can complicate mathematical modeling efforts. For instance, this is seen in the modeling of electrorheological fluids whose viscosity changes under the electric field. Although these fluids are well understood experimentally, a comprehensive theoretical model has deficient aspects. Electrorheological fluids are treated as non-Newtonian in fluid dynamics, and the challenges they reveal have contributed to the development of variable exponent Lebesgue and Sobolev spaces, an area of study that has grown significantly since the 1990s.

Variable Lebesgue spaces have also been used to model the behavior of other physical phenomena, such as thermistor problems, fluid flow in porous media and magnetostatics. In image processing, these spaces are applied to obtain a smoother image through the interpolation technique using the variable exponent. Extensive mathematical literature is available on these topics. Additionally, considerable research has focused on the problems - such as classical Riemann boundary value problems, Dirichlet problems, Dirichlet-Neumann problems- in variable exponent Lebesgue spaces which present much more sensitivity compared to the classical Lebesgue spaces.

The concept of variable exponent spaces was introduced firstly by Orlicz in 1930 [145]. Over the last 20-30 years, these spaces have been actively investigated by mathematical research groups in various locations, Sapporo (Japan), Voronezh and Mahachkala (Russia), Leiden (Netherlands), Tbilisi (Georgia, country), Algarve (Portugal), Baku (Azerbaijan), Balikesir and Istanbul (Turkiye).

Research on fundamental problems in variable exponent Lebesgue spaces has spanned areas (see the monographs [1, 106, 148] and references therein) such as potential theory, maximal and singular integral operators theory, among others. Issues regarding the basicity of exponential systems, trigonometric systems and uniform boundedness of certain convolution operator families in variable exponent Lebesgue spaces have been studied in [96]- [99], [137] and [149], [156].

The foundational properties of variable exponent Lebesgue spaces in relation to approximation theory were addressed by Sharapudinov in his monograph [151]: Sharapudinov I.I.: Some Questions in Approximation Theory for Lebesgue Spaces with Variable Exponent, Itogi Nauki. Yug Rossii. Mat. Monograf. Vol. 5, Southern Mathematical Institute of the Vladikavkaz Scientific Center of the Russian Academy of Sciences and Republic of North Ossetia-Alania, Vladikavkaz 2012, p.267 (Russian). Sharapudinov examined approximation by trigonometric polynomials, the completeness problem of trigonometric systems, and Tsenov's minimization problems.

Numerous researchers continue to investigate approximation theory in variable exponent spaces (see, e.g., [150]- [8], [110]- [112], [90]- [93], [116]- [136], [132], [157]- [159], [156]). This paper provides a review of results in approximation theory, particularly for variable exponent Lebesgue spaces defined on the intervals of the real line and Smirnov classes of analytic functions defined on simply connected domains of the complex plane. Similar issues in classical Lebesgue spaces and Smirnov classes have been discussed, for example in the monographs [163], [155], [153], [154], [104] and in the papers [160], [94], [95], [138], [100], [112]- [115], [131], [162]. This review aims to thoroughly examine fundamental problems in approximation theory and constructive approximation (e.g., the direct and inverse theorems, the constructive characterization problems, the approximation properties of certain summation methods) in variable spaces.

The specific objectives concerning variable exponent Lebesgue and Sobolev spaces, as well as Smirnov classes of analytic functions in this review, are as follows:

a) To investigate the possibility of approximation,

b) to develop methods for constructing approximation polynomials on the real line and the complex plane,

c) to define new moduli of smoothness and study their fundamental properties,

d) to prove direct and inverse theorems in approximation theory,

e) to investigate the approximation properties of various summation methods,

f) to define the Lipschitz classes of functions and obtain their constructive characterizations.

We will draw upon existing studies in mathematical literature and our previous work. First, results for variable exponent spaces on intervals of the real axis will be presented, followed by methods for extending these results to variable exponent spaces on domains of the complex plane. To address the above processes, we employ methods based on the properties of variable exponent spaces, as well as traditional approaches used in classical Lebesgue and Smirnov spaces. Specifically, the results obtained for the complex plane are based significantly on recent findings regarding the boundedness of singular and maximal operators in variable exponent spaces.

While this study does not claim to encompass all findings in the approximation theory within variable exponent spaces, we hope it will partially address existing gaps in the field.

Throughout this work by $c(\cdot), c_1(\cdot), c_2(\cdot),..., c(\cdot,\cdot), c_1(\cdot,\cdot), c_2(\cdot,\cdot),...$ we denote the constants (which can be different in different relations) depending only on the parameters given in the corresponding brackets.

2. Approximation in variable exponent Lebesgue spaces

The main results of this section are based on the authors' works [123] and [125]- [127].

Let $T := [0, 2\pi]$ and $p(\cdot) : T \to [0, \infty)$ be a Lebesgue measurable 2π periodic function such that

$$1 \leq p_- := \operatorname*{ess\ inf}_{x \in T} p(x) \leq \operatorname*{ess\ sup}_{x \in T} p(x) := p^+ < \infty. \tag{3.1}$$

In addition to this requirement, if

$$|p(x) - p(y)| \ln \frac{2\pi}{|x-y|} \leq d, \ \forall x, y \in [0, 2\pi] \tag{3.2}$$

with a positive constant d, then we say that $p(\cdot) \in \mathscr{P}(T)$. We also define $\mathscr{P}_0(T) := \{p(\cdot) \in \mathscr{P}(T) : p_- > 1\}$.

The variable exponent Lebesgue space $L^{p(\cdot)}(T)$ is defined as the set of all Lebesgue measurable 2π periodic functions f such that

$$\rho_{p(\cdot)}(f) := \int_0^{2\pi} |f(x)|^{p(x)} \, dx < \infty.$$

Equipped with the norm

$$\|f\|_{p(\cdot)} = \inf \{ \lambda > 0 : \rho_{p(\cdot)}(f/\lambda) \leq 1 \},$$

it becomes a Banach space.

Note that approximation problems in these spaces have not been widely investigated. Meanwhile, some of the fundamental problems of approximation theory in variable exponent Lebesgue spaces of periodic and non-periodic functions defined on intervals of the real line were studied and solved by Sharapudinov (detailed information can be found in his monograph [151]). In particular, Sharapudinov solved several qualitative approximation problems, identifying conditions under which approximation is feasible. It was proven that the Haar system forms a basis for variable exponent space $L^{p(\cdot)}[0,1]$ if and only if the variable exponent $p(\cdot)$ satisfies the condition (3.2). It was also proven that the trigonometric system $\left\{e^{ikx}\right\}_{k\in\mathbb{Z}}$ is a basis for the space $L^{p(\cdot)}[0,2\pi]$ if and only if the 2π periodic function $p(\cdot)$ satisfies the same condition (3.2).

One of the main problems in the approximation theory is defining correctly modulus of smoothness which provides us a better tool to deal with the rate of best approximation, the inverse theorems and other similar problems. Detailed information about various moduli of smoothness considered in mathematical literature can be found in the monograph [107] and in the papers [146] and [108]. The classical modulus of smoothness constructed by using the shift operator $f(\cdot+h)$ has proven to be very useful for problems of the type mentioned above in classical Lebesgue spaces. But it is a fact that $L^{p(\cdot)}(T)$ is noninvariant with respect to the usual shift operator $f(\cdot+h)$ in general [152, p 146]. Nevertheless, the Steklov mean value operator

$$\sigma_h(f) := \frac{1}{h}\int_0^h f(x+t)\,dt, \quad h>0$$

is bounded in $L^{p(\cdot)}(T)$, which follows from the boundedness of the Hardy-Littlewood maximal operator in $L^{p(\cdot)}(T)$, $p(\cdot)\in\mathscr{P}_0(T)$, showed in [105]. By using this result, the *first order modulus of smoothness*

$$\Omega_{p(\cdot)}(f,\delta) := \sup_{0<h\leq\delta}\left\|\frac{1}{h}\int_0^h |f(\cdot)-f(\cdot+t)|\,dt\right\|_{p(\cdot)}$$

was constructed in [110], and in terms of this modulus the direct theorem of approximation theory in $L^{p(\cdot)}(T)$ when $p(\cdot)\in\mathscr{P}_0(T)$ and some results on approximation by the Nörlund means of Fourier series in $L^{p(\cdot)}(T)$ were obtained. Similar results under the condition $p(\cdot)\in\mathscr{P}_0(T)$ using some other modulus of smoothness were proved in [116], [90]- [93]. In more general cases, i.e. in the case of $p(\cdot)\in\mathscr{P}(T)\supset\mathscr{P}_0(T)$, introducing the modulus

$$\Omega(f,\delta)_{p(\cdot)} := \sup_{0<h\leq\delta}\left\|\frac{1}{h}\int_0^h [f(\cdot)-f(\cdot+t)]\,dt\right\|_{p(\cdot)}$$

Sharapudinov proved the direct and inverse theorems in [152]. Later, defining the rth modulus of smoothness $\Omega_r(f,\delta)_{p(\cdot)}$, $r\in\mathbb{N}$, $p(\cdot)\in\mathscr{P}(T)$, one general inverse theorem in terms of this modulus which generalizes the inverse theorem obtained in [152] was proved in [121].

For the formulation of our main results, first of all, we define above mentioned rth $(r=1,2,...)$ *modulus of smoothness* $\Omega_r(f,\delta)_{p(\cdot)}$ in $L^{p(\cdot)}(T)$, $p(\cdot)\in\mathscr{P}(T)$, obtain its important properties and later prove direct and inverse theorems in $L^{p(\cdot)}(T)$.

2.1. Modulus of smoothness in $L^{p(\cdot)}(T)$ and Steklov means

Definition 2.1. *Let* $f\in L^{p(\cdot)}(T)$ *with* $p(\cdot)\in\mathscr{P}(T)$ *and*

$$\Delta_t^r f(x) := \sum_{s=0}^{r}(-1)^{r+s}\binom{r}{s} f(x+st), \qquad r=1,2,\dots .$$

We define the r-th modulus of smoothness as

$$\Omega_r(f,\delta)_{p(\cdot)} := \sup_{0<h\le\delta}\left\|\frac{1}{h}\int_0^h \Delta_t^r f dt\right\|_{p(\cdot)}, \quad \delta>0.$$

It is easy to show that in the case of $p(\cdot)=$ constant this modulus is equivalent to the classical modulus of smoothness defined as $\Omega_r(f,\delta):=\sup_{|t|\le\delta}\|\Delta_t^r f(x)\|_p$.

Subadditivity property of $\Omega_r(f,\delta)_{p(\cdot)}$ immediately follows from *Definition 2.1*; that is for $f,g\in L^{p(\cdot)}(T)$ we have

$$\Omega_r(f+g,\delta)_{p(\cdot)} \le \Omega_r(f,\delta)_{p(\cdot)} + \Omega_r(g,\delta)_{p(\cdot)}. \tag{3.3}$$

Let $\lambda,\gamma>0$, and $|\tau|\le\pi/\lambda^\gamma$. We consider the Steklov operator

$$\mathscr{S}_{\lambda,\tau}f := (\mathscr{S}_{\lambda,\tau}f)(x) := \lambda\int_{x+\tau-1/2\lambda}^{x+\tau+1/2\lambda} f(t)\,dt.$$

The following lemma was proved in [150].

Lemma 2.1 ([150]). *Let* $p(\cdot)\in\mathscr{P}(T)$ *and* $0<\gamma\le 1$. *Then the family of the Steklov operators* $S_{\lambda,\tau}(f)$ *is uniformly bounded in* $L^{p(\cdot)}(T)$ *for* $1\le\lambda<\infty$, $|\tau|\le\pi/\lambda^\gamma$, *i.e., there exists a positive constant* $c(p)$ *such that*

$$\|\mathscr{S}_{\lambda,\tau}f\|_{p(\cdot)} \le c(p)\,\|f\|_{p(\cdot)}, \quad 1\le\lambda<\infty, \ |\tau|\le\pi/\lambda^\gamma.$$

The lemma below shows that the modulus $\Omega_r(f,\delta)_{p(\cdot)}$ is well defined.

Lemma 2.2. *Let $p(\cdot)\in\mathscr{P}(T)$ and $r\in\mathbb{N}$. Then there exists a positive constant $c(p,r)$ such that*

$$\Omega_r(f,\delta)_{p(\cdot)}\le c(p,r)\|f\|_{p(\cdot)}$$

for every $f\in L^{p(\cdot)}(T)$ and $\delta>0$.

Proof. Since for any positive integer s with $0<sh\le 1$

$$\frac{1}{h}\int_0^h f(x+st)\,dt=\frac{1}{sh}\int_x^{x+sh} f(u)\,du=\frac{1}{sh}\int_{x+sh/2-sh/2}^{x+sh/2+sh/2} f(u)\,du=\left(\mathscr{S}_{\frac{1}{sh},\frac{sh}{2}}f\right)(x),$$

denoting $\lambda:=1/(sh)$, $\tau:=sh/2$ and applying Lemma 2.1 we have

$$\left\|\frac{1}{h}\int_0^h f(\cdot+st)\,dt\right\|_{p(\cdot)}=\left\|\mathscr{S}_{\frac{1}{sh},\frac{sh}{2}}f\right\|_{p(\cdot)}\le c(p)\|f\|_{p(\cdot)}.$$

Hence

$$\begin{aligned}\left\|\frac{1}{h}\int_0^h \Delta_t^r f dt\right\|_{p(\cdot)} &= \left\|\frac{1}{h}\int_0^h \sum_{s=0}^{r}(-1)^{r+s}\binom{r}{s}f(\cdot+st)\,dt\right\|_{p(\cdot)}\\ &\le c_1(r)\sum_{s=0}^{r}\left\|\frac{1}{h}\int_0^h f(\cdot+st)\,dt\right\|_{p(\cdot)}\\ &\le c(p,r)\|f\|_{p(\cdot)},\end{aligned}$$

which implies that

$$\Omega_r(f,\delta)_{p(\cdot)}\le c(p,r)\|f\|_{p(\cdot)}.$$

□

Lemma 2.3. *If $f\in L^{p(\cdot)}(T)$, $p(\cdot)\in\mathscr{P}(T)$, then $\lim\limits_{\delta\to 0}\Omega_r(f,\delta)_{p(\cdot)}=0$ for every positive integer r.*

Proof. At first we suppose that $f\in L^{p(\cdot)}(T)$ is a continuous function. Then for any $\varepsilon>0$ there is a number $\delta(\varepsilon)>0$ such that $|f(x+mt)-f(x+(m+1)t)|$ $\le\varepsilon/\left\{[2\pi]^{1/p_-}\left(2^{r-1}\right)\right\}$ for $0<t\le h\le\delta$ and $m=0,1,2,...,r$. Hence

$$\int_{-\pi}^{\pi}\left|\frac{1}{h}\int_0^h \Delta_t^r f(x)\,dt/\varepsilon\right|^{p(x)}dx=\int_{-\pi}^{\pi}\left|\frac{1}{h}\int_0^h\frac{\sum_{s=0}^{r}(-1)^{r+s}\binom{r}{s}f(x+st)}{\varepsilon}dt\right|^{p(x)}dx$$

$$\leq \int_{-\pi}^{\pi} \left(\frac{1}{h} \int_0^h \frac{\left| \sum_{s=0}^{r} (-1)^{r+s} \binom{r}{s} f(x+st) \right|}{\varepsilon} dt \right)^{p(x)} dx$$

$$= \int_{-\pi}^{\pi} \left(\frac{1}{h} \int_0^h \frac{\left| \sum\limits_{m=0}^{r-1} (-1)^{r+m} \binom{r-1}{m} [f(x+mt) - f(x+(m+1)t)] \right|}{\varepsilon} dt \right)^{p(x)} dx$$

$$\leq \int_{-\pi}^{\pi} \left(\frac{1}{h} \int_0^h \frac{\sum\limits_{m=0}^{r-1} \binom{r-1}{m} |f(x+mt) - f(x+(m+1)t)|}{\varepsilon} dt \right)^{p(x)} dx$$

$$\leq \int_{-\pi}^{\pi} \left(\frac{1}{h} \int_0^h \frac{\varepsilon \sum\limits_{m=0}^{r-1} \binom{r-1}{m}}{[2\pi]^{1/p_-} (2^{r-1}) \varepsilon} dt \right)^{p(x)} dx$$

$$= \int_{-\pi}^{\pi} \left(\frac{1}{h} \int_0^h \frac{(2^{r-1})}{[2\pi]^{1/p_-} (2^{r-1})} dt \right)^{p(x)} dx$$

$$\leq \int_{-\pi}^{\pi} \left(\frac{1}{h} \int_0^h \frac{dt}{[2\pi]^{1/p_-}} \right)^{p_-} dx = 1.$$

Therefore,

$$\left\| \frac{1}{h} \int_0^h \Delta_t^r (f) \right\|_{p(\cdot)} \leq \varepsilon$$

which implies that $\Omega_r (f, \delta)_{p(\cdot)} \leq \varepsilon$.

If $f \in L^{p(\cdot)}(T)$ is not continuous on T, then by density of the set of continuous functions [151, pp. 145-146] in $L^{p(\cdot)}(T)$, for any $\varepsilon > 0$ there exists a 2π periodic continuous function g with $\|f-g\|_{p(\cdot)} \leq \varepsilon$ and a number $\delta(\varepsilon) > 0$ such that $\Omega_r (g, \delta)_{p(\cdot)} \leq \varepsilon$ for every $\delta < \delta(\varepsilon)$. Hence by (3.3) and Lemma 2.2

$$\begin{aligned} \Omega_r (f, \delta)_{p(\cdot)} &\leq \Omega_r (f-g, \delta)_{p(\cdot)} + \Omega_r (g, \delta)_{p(\cdot)} \\ &\leq c(p,r) \|f-g\|_{p(\cdot)} + \varepsilon \\ &\leq [c(p,r)+1] \varepsilon, \end{aligned}$$

which implies the required relation $\lim\limits_{\delta \to 0} \Omega_r (f, \delta)_{p(\cdot)} = 0$. □

Let

$$W_r^{p(\cdot)}(T) := \left\{ f : f^{(r-1)} \text{ is absolutely continuous and } f^{(r)} \in L^{p(\cdot)}(T) \right\}$$

$r = 1, 2, ...,$ be the variable exponent Sobolev space. We will prove the lemma below, which can be used for the proof of inverse theorems in $W_r^{p(\cdot)}(T)$.

Lemma 2.4. *Let $p(\cdot) \in \mathscr{P}(T)$. Then there exists a positive constant $c(p,r)$ such that for any $r \in \mathbb{N}$, $\delta > 0$ and for any function $f \in W_r^{p(\cdot)}(T)$ the inequality*

$$\Omega_r(f,\delta)_{p(\cdot)} \leq c(p,r)\delta^r \left\| f^{(r)} \right\|_{p(\cdot)}$$

holds.

Proof. Since

$$\Delta_t^r f(x) = \int_0^t \int_0^t ... \int_0^t f^{(r)}(x + t_1 + ... + t_r)\, dt_1 ... dt_r,$$

applying r times the generalized Minkowski inequality we have

$$\left\| \frac{1}{h} \int_0^h \Delta_t^r f dt \right\|_{p(\cdot)} \leq c_1(p) \frac{1}{h} \int_0^h \|\Delta_t^r f\|_{p(\cdot)}\, dt$$

$$\begin{aligned}
&\leq c_1(p)h^r \frac{1}{h^{r+1}} \int_0^h \left\| \int_0^t \cdots \int_0^t f^{(r)}(\cdot + t_1 + \ldots + t_r)\, dt_1 \ldots dt_r \right\|_{p(\cdot)} dt \\
&\leq c_1(p)h^r \frac{1}{h} \int_0^h \left\| \frac{1}{h} \int_0^t \left| \frac{1}{h^{r-1}} \int_0^t \cdots \int_0^t f^{(r)}(\cdot + t_1 + \ldots + t_r)\, dt_1 \ldots dt_{r-1} \right| dt_r \right\|_{p(\cdot)} dt \\
&= c_1(p)h^r \frac{1}{h} \int_0^h \left\| \frac{1}{h} \int_0^h \left| \frac{1}{h^{r-1}} \int_0^t \cdots \int_0^t f^{(r)}(\cdot + t_1 + \ldots + t_r)\, dt_1 \ldots dt_{r-1} \right| dt_r \right\|_{p(\cdot)} dt \\
&\leq c_2(p)h^r \frac{1}{h} \int_0^h \left\| \frac{1}{h^{r-1}} \int_0^t \cdots \int_0^t f^{(r)}(\cdot + t_1 + \ldots + t_{r-1})\, dt_1 \ldots dt_{r-1} \right\|_{p(\cdot)} dt \\
&\leq \ldots \leq c_3(p,r)h^r \frac{1}{h} \int_0^h \left\| \left\{ \frac{1}{h} \int_0^h \left| f^{(r)}(\cdot + t_1) \right| dt_1 \right\} \right\|_{p(\cdot)} dt \\
&\leq c_4(p,r)h^r \left\| f^{(r)} \right\|_{p(\cdot)} \frac{1}{h} \int_0^h dt = c_4(p,r)h^r \left\| f^{(r)} \right\|_{p(\cdot)},
\end{aligned}$$

and taking here the supremum we obtain the inequality

$$\Omega_r(f,\delta)_{p(\cdot)} \leq c(p,r)\delta^r \left\| f^{(r)} \right\|_{p(\cdot)}.$$

□

For $f \in L^{p(\cdot)}(T)$ and $\delta > 0$ we define the Steklov mean value function

$$\begin{aligned}
f_{r,\delta}(x) &= \frac{2}{\delta} \int_{\delta/2}^{\delta} \left\{ \frac{1}{h^r} \sum_{s=0}^{r-1} (-1)^{r+s+1} \binom{r}{s} \int_0^h \cdots \right. \\
&\left. \cdots \int_0^h f\left(x + \frac{r-s}{r}[t_1 + \ldots + t_r] \right) dt_1 \ldots dt_r \right\} dh, \qquad (3.4)
\end{aligned}$$

which plays a crucial role in this work.

Lemma 2.5. *If $f \in L^{p(\cdot)}(T)$, $p(\cdot) \in \mathscr{P}(T)$, then $f_{r,\delta} \in W_r^{p(\cdot)}(T)$ for $\delta > 0$ and $r \in \mathbb{N}$.*

Proof. Differentiating $r-1$ times the terms under the sum in (3.4) and setting $t := \frac{r-s}{r} t_r$ we see that

$$\left\{ \int_0^h \cdots \int_0^h f\left(x + \frac{r-s}{r}\left[t_1 + \ldots + t_r\right]\right) dt_1 \ldots dt_r \right\}^{(r-1)}$$

$$= \left\{ \int_0^h \left(\frac{r}{r-s}\right)^{r-1} \sum_{m=0}^{r-1} \binom{r-1}{m} (-1)^{r+m} f\left(x + \frac{r-s}{r} t_r + m\frac{r-s}{r} h\right) dt_r \right\}$$

$$= \int_0^h \left(\frac{r}{r-s}\right)^{r-1} \Delta^{r-1}_{\frac{r-s}{r}h} f\left(x + \frac{r-s}{r} t_r\right) dt_r$$

$$= \int_0^{\frac{r-s}{r}h} \left(\frac{r}{r-s}\right)^{r} \Delta^{r-1}_{\frac{r-s}{r}h} f(x+t)\, dt$$

and then via (3.4)

$$f^{(r-1)}_{r,\delta}(x) = \frac{2}{\delta} \int_{\delta/2}^{\delta} \frac{1}{h^r} \left\{ \sum_{s=0}^{r-1} (-1)^{r+s+1} \binom{r}{s} \int_0^{\frac{r-s}{r}h} \left(\frac{r}{r-s}\right)^{r} \Delta^{r-1}_{\frac{r-s}{r}h} f(x+t)\, dt \right\} dh$$

$$= \frac{2}{\delta} \int_{\delta/2}^{\delta} \frac{1}{h^r} \left\{ \sum_{s=0}^{r-1} \int_x^{x+\frac{r-s}{r}h} (-1)^{r+s+1} \binom{r}{s} \left(\frac{r}{r-s}\right)^{r} \Delta^{r-1}_{\frac{r-s}{r}h} f(t)\, dt \right\} dh. \quad (3.5)$$

Since the Steklov mean value function $f^{(r-1)}_{r,\delta}$ is a definite Lebesgue integral on $[0, 2\pi]$, its absolute continuity on $[0, 2\pi]$ can be shown in a standard way. It remains to prove the embedding $f^{(r)}_{r,\delta} \in L^{p(\cdot)}(T)$. Differentiating the relation (3.5) we obtain

$$f^{(r)}_{r,\delta}(x) = \frac{2}{\delta} \int_{\delta/2}^{\delta} \frac{1}{h^r} \left(\sum_{s=0}^{r-1} (-1)^{r+s+1} \binom{r}{s} \left(\frac{r}{r-s}\right)^{r} \Delta^{r}_{\frac{r-s}{r}h} f(x) \right) dh$$

and denoting $t := \dfrac{r-s}{r}h$ we have

$$\begin{aligned}
\left|f_{r,\delta}^{(r)}(x)\right| &\leq \frac{2^{r+1}}{\delta^r}\sum_{s=0}^{r-1}\binom{r}{s}\left(\frac{r}{r-s}\right)^r\left|\frac{1}{\delta}\int_{\delta/2}^{\delta}\Delta^r_{\frac{r-s}{r}h}f(x)\,dh\right| \\
&= \frac{2^{r+1}}{\delta^r}\sum_{s=0}^{r-1}\binom{r}{s}\left(\frac{r}{r-s}\right)^r\left|\frac{1}{\frac{r-s}{r}\delta}\int_{\frac{r-s}{r}(\delta/2)}^{\frac{r-s}{r}\delta}\Delta^r_t f(x)\,dt\right| \\
&\leq \frac{2^{r+1}}{\delta^r}\sum_{s=0}^{r-1}\binom{r}{s}\left(\frac{r}{r-s}\right)^r\left\{\left|\frac{1}{\frac{r-s}{r}\delta}\int_{0}^{\frac{r-s}{r}\delta}\Delta^r_t f(x)\,dt\right|\right. \\
&\quad \left.+\left|\frac{1}{\frac{r-s}{r}\delta}\int_{0}^{\frac{r-s}{r}(\delta/2)}\Delta^r_t f(x)\,dt\right|\right\},
\end{aligned}$$

which by Lemma 2.2 implies the inequality

$$\left\|f_{r,\delta}^{(r)}\right\|_{p(\cdot)} \leq 2c(r)\delta^{-r}\Omega_r(f,\delta)_{p(\cdot)} \leq c_5(p,r)\|f\|_{p(\cdot)}. \tag{3.6}$$

Since $f \in L^{p(\cdot)}(T)$, the relation (3.6) means that $f_{r,\delta}^{(r)} \in L^{p(\cdot)}(T)$. □

For $f \in L^{p(\cdot)}(T)$ we define the best approximation number

$$E_n(f)_{p(\cdot)} := \inf\left\{\|f-T_n\|_{p(\cdot)} : T_n \in \Pi_n\right\}$$

in the class Π_n of the trigonometric polynomials of degree not exceeding n.

In the variable exponent Sobolev space $W_k^{p(\cdot)}(T)$, using *Consequence 2.1* in [8] and the boundedness of $\Omega(f,\delta)_{p(\cdot)}$, we have the inequality

$$E_n(f)_{p(\cdot)} \leq \frac{c(p)}{n^k}\left\|f^{(k)}\right\|_{p(\cdot)}, \tag{3.7}$$

which implies, by the standard way, the estimate

$$E_n(f)_{p(\cdot)} \leq \frac{c(p)}{n^k}E_n\left(f^{(k)}\right)_{p(\cdot)}. \tag{3.8}$$

2.2. Direct and Inverse Theorems

In this subsection, we will formulate and prove some direct and inverse theorems in variable exponent Lebesgue spaces $L^{p(\cdot)}(T)$.

The Main direct theorem obtained in $L^{p(\cdot)}(T)$ can be formulated as follows.

Theorem 2.6. *Let $p(\cdot) \in \mathscr{P}(T)$, $r \in \mathbb{N}$. Then there exists a positive constant $c(p,r)$ such that for every $f \in L^{p(\cdot)}(T)$ and $n \in \mathbb{N}$ the inequality*

$$E_n(f)_{p(\cdot)} \leq c(p,r)\Omega_r(f,1/n)_{p(\cdot)}$$

holds.

Proof. Let $f \in L^{p(\cdot)}(T)$, $p(\cdot) \in \mathscr{P}(T)$. For $\delta > 0$ and $r = 1,2,...$, after some necessary simplifications, we have

$$\left|f_{r,\delta}(x) - f(x)\right| = \frac{2}{\delta}\left|\int\limits_{\delta/2}^{\delta}\left\{\frac{1}{h^r}\int\limits_0^h \cdots \int\limits_0^h \Delta^r_{\frac{t_1+\ldots+t_r}{r}} f(x)\, dt_1 \ldots dt_r\right\} dh\right|$$

and then by the generalized Minkowski inequality

$$\left\|f_{r,\delta} - f\right\|_{p(\cdot)} \leq c_6(p,r)\frac{2}{\delta}\int\limits_{\delta/2}^{\delta}\left\{\frac{1}{h^{r-1}}\int\limits_0^h \cdots \int\limits_0^h \left\|\frac{1}{h}\int\limits_0^h \Delta^r_{\frac{t_1+\ldots+t_r}{r}} f dt_1\right\|_{p(\cdot)} dt_2 \ldots dt_r\right\} dh$$

$$= c_6(p,r)\frac{2}{\delta}\int\limits_{\delta/2}^{\delta}\left\{\frac{1}{h^{r-1}}\int\limits_0^h \cdots \int\limits_0^h \left\|\frac{1}{h}\int\limits_{t_2+\ldots+t_r}^{h+t_2+\ldots+t_r} \Delta^r_{\frac{t}{r}} f dt\right\|_{p(\cdot)} dt_2 \ldots dt_r\right\} dh. \quad (3.9)$$

Since

$$\left\|\frac{1}{h}\int\limits_{t_2+\ldots+t_r}^{h+t_2+\ldots+t_r} \Delta^r_{\frac{t}{r}} f dt\right\|_{p(\cdot)} = \left\|\frac{1}{h}\left(\int\limits_0^{h+t_2+\ldots+t_r} \Delta^r_{\frac{t}{r}} f dt - \int\limits_0^{t_2+\ldots+t_r} \Delta^r_{\frac{t}{r}} f dt\right)\right\|_{p(\cdot)}$$

$$\leq \left\| \frac{1}{(h+t_2+...+t_r)/r} \int\limits_0^{(h+t_2+...+t_r)/r} \Delta_t^r f dt \right\|_{p(\cdot)}$$

$$+ \left\| \frac{1}{(t_2+...+t_r)/r} \int\limits_0^{(t_2+...+t_r)/r} \Delta_t^r f dt \right\|_{p(\cdot)}$$

$$\leq \sup_{(h+t_2+...+t_r)/r\leq\delta} \left\| \frac{1}{(h+t_2+...+t_r)/r} \int\limits_0^{(h+t_2+...+t_r)/r} \Delta_t^r f dt \right\|_{p(\cdot)}$$

$$+ \sup_{(t_2+...+t_r)/r\leq\delta} \left\| \frac{1}{(t_2+...+t_r)/r} \int\limits_0^{(t_2+...+t_r)/r} \Delta_t^r f dt \right\|_{p(\cdot)}$$

$$= \Omega_r(f,\delta)_{p(\cdot)} + \Omega_r(f,\delta)_{p(\cdot)} = 2\Omega_r(f,\delta)_{p(\cdot)}, \tag{3.10}$$

combining (3.9) and (3.10) we have

$$\left\| f_{r,\delta} - f \right\|_{p(\cdot)} \leq c_7(p,r)\frac{2}{\delta}\int\limits_{\delta/2}^{\delta} \left\{ \frac{1}{h^{r-1}} \int\limits_0^h ... \int\limits_0^h \Omega_r(f,\delta)_{p(\cdot)} dt_2...dt_r \right\} dh$$

$$\leq c_7(p,r)\Omega_r(f,\delta)_{p(\cdot)} \frac{2}{\delta}\int\limits_{\delta/2}^{\delta} dh = c_7(p,r)\Omega_r(f,\delta)_{p(\cdot)}. \tag{3.11}$$

Since by (3.7)

$$E_n\left(f_{r,1/n}\right)_{p(\cdot)} \leq \frac{c(p)}{n^r}\left\| f_{r,1/n}^{(r)} \right\|_{p(\cdot)},$$

using the relations (3.11) and (3.6), respectively, we conclude that

$$E_n(f)_{p(\cdot)} \leq E_n\left(f - f_{r,1/n}\right)_{p(\cdot)} + E_n\left(f_{r,1/n}\right)_{p(\cdot)}$$

$$\leq \left\| f_{r,1/n} - f \right\|_{p(\cdot)} + \frac{c(p)}{n^r}\left\| f_{r,1/n}^{(r)} \right\|_{p(\cdot)}$$

$$\leq c_8(p,r)\Omega_r(f,1/n)_{p(\cdot)} + \frac{c_9(p,r)}{n^r} n^r \Omega_r(f,1/n)_{p(\cdot)}$$

$$\leq c(p,r)\Omega_r(f,1/n)_{p(\cdot)}.$$

□

Theorem 2.6 in the case of $r = 1$ was proved in [110] (when $p(\cdot) \in \mathscr{P}_0(T)$) and in [152] (when $p(\cdot) \in \mathscr{P}(T)$).

Combining the estimate (3.8) with Theorem 2.6, we have

Corollary 2.7. *Let $p(\cdot) \in \mathscr{P}(T)$, $k \in \mathbb{N}$. Then there exists a positive constant $c(p,r)$ such that for every $f \in W_k^{p(\cdot)}(T)$ and $n \in \mathbb{N}$ the inequality*

$$E_n(f)_{p(\cdot)} \leq \frac{c(p,r)}{n^k}\Omega_r\left(f^{(k)}, 1/n\right)_{p(\cdot)}$$

holds.

In the case of $r = 1$ *Corollary 2.7* was obtained in [8].

The inverse theorem can be formulated as follows:

Theorem 2.8. *Let $p(\cdot) \in \mathscr{P}(T)$, $r \in \mathbb{N}$. Then there exists a positive constant $c(p,r)$ such that for every $f \in L^{p(\cdot)}(T)$ and $n \in \mathbb{N}$ the inequality*

$$\Omega_r(f, 1/n)_{p(\cdot)} \leq \frac{c(p,r)}{n^r}\sum_{k=0}^{n}(k+1)^{r-1}E_k(f)_{p(\cdot)}$$

holds.

Proof. Let T_n be the best approximation trigonometric polynomial for $f \in L^{p(\cdot)}(T)$, $p(\cdot) \in \mathscr{P}(T)$. Let also $m \in \mathbb{N}$ be the number such that $2^m \leq n < 2^{m+1}$. Since

$$\Omega_r(f, 1/n)_{p(\cdot)} \leq \Omega_r(f - T_{2^{m+1}}, 1/n)_{p(\cdot)} + \Omega_r(T_{2^{m+1}}, 1/n)_{p(\cdot)}, \tag{3.12}$$

using the inequality [104, p. 209]

$$2^{(\nu+1)r}E_{2^\nu}(f)_{p(\cdot)} \leq 2^{2r}\sum_{k=2^{\nu-1}+1}^{2^\nu} k^{r-1}E_k(f)_{p(\cdot)}, \tag{3.13}$$

we have

$$\Omega_r(f - T_{2^{m+1}}, 1/n)_{p(\cdot)} \leq c(p,r)\left\|f - T_{2^{m+1}}\right\|_{p(\cdot)} = c(p,r)E_{2^{m+1}}(f)_{p(\cdot)}$$

$$\leq c(p,r)\frac{2^{(m+1)r}}{n^r}E_{2^m}(f)_{p(\cdot)} \leq \frac{c(p,r)}{n^r}2^{2r}\sum_{k=2^{m-1}+1}^{2^m} k^{r-1}E_k(f)_{p(\cdot)}. \tag{3.14}$$

On the other hand, applying Lemma 2.4 and the Bernstein inequality for trigonometric polynomials:

$$\left\|T_n'\right\|_{p(\cdot)} \leq c(p)n\left\|T_n\right\|_{p(\cdot)},$$

proved in [152], and (3.13), we get

$$\Omega_r\left(T_{2^{m+1}},1/n\right)_{p(\cdot)} \leq \frac{c(p,r)}{n^r}\left\|T^{(r)}_{2^{m+1}}\right\|_{p(\cdot)}$$

$$\begin{aligned}
&= \frac{c(p,r)}{n^r}\left\|T_1^{(r)}+\sum_{\nu=0}^{m}\left(T^{(r)}_{2^{\nu+1}}-T^{(r)}_{2^{\nu}}\right)\right\|_{p(\cdot)} \\
&\leq \frac{c(p,r)}{n^r}\left(\left\|T_1^{(r)}\right\|_{p(\cdot)}+\left\|\sum_{\nu=0}^{m}\left(T^{(r)}_{2^{\nu+1}}-T^{(r)}_{2^{\nu}}\right)\right\|_{p(\cdot)}\right) \\
&\leq \frac{c_{10}(p,r)}{n^r}\left(\|T_1\|_{p(\cdot)}+\sum_{\nu=0}^{m}2^{(\nu+1)r}\left\|\left(T_{2^{\nu+1}}-T_{2^{\nu}}\right)\right\|_{p(\cdot)}\right) \\
&\leq \frac{c_{11}(p,r)}{n^r}\left(E_0(f)_{p(\cdot)}+\sum_{\nu=0}^{m}2^{(\nu+1)r}E_{2^{\nu}}(f)_{p(\cdot)}\right) \\
&= \frac{c_{11}(p,r)}{n^r}\left(E_0(f)_{p(\cdot)}+2^rE_1(f)_{p(\cdot)}+\sum_{\nu=1}^{m}2^{(\nu+1)r}E_{2^{\nu}}(f)_{p(\cdot)}\right) \\
&\leq \frac{c_{11}(p,r)}{n^r}\left(E_0(f)_{p(\cdot)}+2^rE_0(f)_{p(\cdot)}+\sum_{\nu=1}^{m}\sum_{k=2^{\nu-1}+1}^{2^{\nu}}k^{r-1}E_k(f)_{p(\cdot)}\right)
\end{aligned}$$

$$\leq \frac{c_{12}(p,r)}{n^r}\left(E_0(f)_{p(\cdot)}+\sum_{k=1}^{2^m}k^{r-1}E_k(f)_{p(\cdot)}\right). \tag{3.15}$$

Combining (3.12), (3.14) and (3.15), we conclude that

$$\begin{aligned}
&\Omega_r(f,1/n)_{p(\cdot)} \\
&\leq \frac{c_{13}(p,r)}{n^r}\left(\sum_{k=2^{m-1}+1}^{2^m}k^{r-1}E_k(f)_{p(\cdot)}+E_0(f)_{p(\cdot)}+\sum_{k=1}^{2^m}k^{r-1}E_k(f)_{p(\cdot)}\right) \\
&\leq \frac{c_{14}(p,r)}{n^r}\left(\sum_{k=1}^{2^m}k^{r-1}E_k(f)_{p(\cdot)}+E_0(f)_{p(\cdot)}\right) \\
&\leq \frac{c(p,r)}{n^r}\left(\sum_{k=0}^{n}(k+1)^{r-1}E_k(f)_{p(\cdot)}\right).
\end{aligned}$$

□

Theorem 2.8 in the case of $r=1$ was proved in [121]. Under the conditions $f\in L^{p(\cdot)}(T)$ and $E_n(f)_{p(\cdot)}\leq cn^{\alpha}$, $\alpha\in(0,1]$, Theorem 2.8 was obtained in [151, pp.167-170].

Theorem 2.8 implies

Corollary 2.9. *If $E_n(f)_{p(\cdot)} = \mathcal{O}\left(n^{-\alpha}\right)$, $\alpha > 0$, then under the conditions of Theorem 2.8,*

$$\Omega_r(f,\delta)_{p(\cdot)} = \begin{cases} \mathcal{O}(\delta^\alpha) & ,r > \alpha \\ \mathcal{O}(\delta^\alpha \log(1/\delta)) & ,r = \alpha \\ \mathcal{O}(\delta^r) & ,r < \alpha. \end{cases}$$

Hence, if we define a generalized Lipschitz class $Lip_{\alpha,r}^{p(\cdot)}(T)$ for $\alpha > 0$ and $r := [\alpha]+1$ ($[\alpha]$ is the integer part of α) as

$$Lip_{\alpha,r}^{p(\cdot)}(T) := \left\{f \in L^{p(\cdot)}(T) : \Omega_r(f,\delta)_{p(\cdot)} = \mathcal{O}(\delta^\alpha),\ \ \delta > 0\right\},$$

then we have

Corollary 2.10. *If $E_n(f)_{p(\cdot)} = \mathcal{O}\left(n^{-\alpha}\right)$, $\alpha > 0$, then under the conditions of Theorem 2.8, $f \in Lip_{\alpha,r}^{p(\cdot)}(T)$.*

On the other hand, from Theorem 2.6 we also get

Corollary 2.11. *If $f \in Lip_{\alpha,r}^{p(\cdot)}(T)$ with $p(\cdot) \in \mathcal{P}(T)$ and some $\alpha > 0$, then $E_n(f)_{p(\cdot)} = \mathcal{O}\left(n^{-\alpha}\right)$.*

Now Corollaries 2.10 and 2.11 give the following constructive characterization of $Lip_{\alpha,r}^{p(\cdot)}(T)$:

Theorem 2.12. *Let $f \in L^{p(\cdot)}(T)$, $p(\cdot) \in \mathcal{P}(T)$, and $\alpha > 0$. The following statements are equivalent:*

$$i)\ f \in Lip_{\alpha,r}^{p(\cdot)}(T),$$

$$ii)\ E_n(f)_{p(\cdot)} = \mathcal{O}\left(n^{-\alpha}\right), n \in \mathbb{N}.$$

Note that when $p(\cdot)$ =*constant*, the classical analogues of Theorems 2.6-2.12, proved in terms of the classical r-th modulus of smoothness constructed via usual shift $f(\cdot + h)$, can be found in the monographs [104, Chapter 7, pp. 201-211, Theorems 2.3, 3.1, 3.3.] and [155, Chapter 5, pp. 325,331].

2.3. Direct and Inverse Theorems in Weighted Case

In this subsection, we prove some direct and inverse theorems of approximation theory in weighted variable exponent Lebesgue spaces $L_{\omega}^{p(\cdot)}(T)$, $p(\cdot) \in \mathscr{P}_0(T)$.

Let ω be a weight function on T, i.e. an almost everywhere positive and Lebesgue integrable function on T.

For a given weight ω, we define the weighted variable exponent Lebesgue space $L_{\omega}^{p(\cdot)}(T)$ as the set of all measurable functions f on T such that $f\omega \in L^{p(\cdot)}(T)$. The norm of $f \in L_{\omega}^{p(\cdot)}(T)$ can be defined as $\|f\|_{p(\cdot),\omega} := \|f\omega\|_{p(\cdot)}$.

Definition 2.2. *We say that $\omega \in A_{p(\cdot)}(T)$ if the inequality*

$$\sup_{I \subset T} |I|^{-1} \|\omega \chi_I\|_{p(\cdot)} \left\|\omega^{-1}\chi_I\right\|_{p'(\cdot)} < \infty, \quad 1/p(\cdot) + 1/p'(\cdot) = 1$$

holds, where $|I|$ is the Lebesgue measure of the interval $I \subset T$ with the characteristic function χ_I.

Let $I \subset T$ be any open interval and

$$Mf(x) := \sup_{I \ni x} \frac{1}{|I|} \int_I |f(t)|\,dt, \quad x \in T,$$

be the Hardy-Littlewood maximal function for f. As follows from [102], if $\omega(\cdot) \in A_{p(\cdot)}(T)$ and $p(\cdot) \in \mathscr{P}_0(T)$, then the maximal operator Mf is bounded in $f \in L_{\omega}^{p(\cdot)}(T)$. In this case, there exists a positive constant $c(p)$ such that the inequality

$$\|M(f)\|_{p(\cdot),\omega} \leq c(p)\,\|f)\|_{p(\cdot),\omega} \tag{3.16}$$

holds.

Definition 2.3. *Let $f \in L_{\omega}^{p(\cdot)}(T)$, $p(\cdot) \in \mathscr{P}_0(T)$ and $\omega(\cdot) \in A_{p(\cdot)}(T)$. We define the modulus of smoothness as*

$$\Omega(f,\delta)_{p(\cdot),\omega} := \sup_{|h| \leq \delta} \left\| \frac{1}{h} \int_0^h (f(x+t) - f(x))\,dt \right\|_{p(\cdot),\omega}, \quad \delta > 0.$$

Clearly $\Omega(f,\delta)_{p(\cdot),\omega}$ is well defined because of $\Omega(f,\delta)_{p(\cdot),\omega} \leq c(p)\,\|f\|_{p(\cdot),\omega}$, by estimate (3.16).

Furthermore, it can be shown easily that if $f, g \in L_{\omega}^{p(\cdot)}(T)$, then

$$\Omega(f+g,\delta)_{p(\cdot),\omega} \leq \Omega(f,\delta)_{p(\cdot),\omega} + \Omega(g,\delta)_{p(\cdot),\omega} \text{ and } \lim_{\delta \to 0} \Omega(f,\delta)_{p(\cdot),\omega} = 0.$$

Let's define the variable exponent weighted Lipschitz class $Lip(\alpha, p(\cdot), \omega)$, $0 < \alpha \leq 1$, as

$$Lip(\alpha, p(\cdot), \omega) := \left\{ f \in L_{\omega}^{p(\cdot)}(T) : \Omega(f, \delta)_{p(\cdot),\omega} = \mathcal{O}(\delta^{\alpha}), \ \ \delta > 0 \right\}.$$

For $f \in L_{\omega}^{p(\cdot)}(T)$ we also define the best approximation number

$$E_n(f)_{p(\cdot),\omega} := \inf \left\{ \|f - T_n\|_{p(\cdot),\omega} : T_n \in \Pi_n \right\}$$

in the class Π_n of the trigonometric polynomials of degree not exceeding n.

Let $W_{\omega}^{p(\cdot),1}(T)$ be the weighted Sobolev space with variable exponent $p(\cdot)$ defined by

$$W_{\omega}^{p(\cdot),1}(T) := \left\{ f : f \text{ is absolutely continuous on } \mathbb{T} \text{ and } f' \in L_{\omega}^{p(\cdot)}(T) \right\}.$$

The proof of the following lemma can be found in [123]:

Lemma 2.13. *If $f \in W_{\omega}^{p(\cdot),1}(T)$, $p(\cdot) \in \mathcal{P}_0(T)$, $\omega \in A_{p(\cdot)}(T)$, then there exists a positive constant $c(p)$ such that the inequality*

$$E_n(f)_{p(\cdot),\omega} = \mathcal{O}\left(\|f'\|_{p(\cdot),\omega} / n \right), \ \ n = 1, 2, ...$$

holds.

Now we can formulate the direct theorem of approximation theory in the weighted variable exponent spaces $L_{\omega}^{p(\cdot)}(T)$.

For simplicity, we consider the results for the first modulus of smoothness.

Theorem 2.14. *If $f \in L_{\omega}^{p(\cdot)}(T)$, $p(\cdot) \in \mathcal{P}_0(T)$, $\omega(\cdot) \in A_{p(\cdot)}(T)$, then*

$$E_n(f)_{p(\cdot),\omega} \leq \mathcal{O}\left(\Omega(f, 1/n)_{p(\cdot),\omega} \right), \ n = 1, 2, \ldots.$$

Proof. We define the Steklov mean operator as

$$f_{\delta}(x) := \frac{2}{\delta} \int_{\delta/2}^{\delta} \left(\frac{1}{h} \int_0^h f(x+t)\, dt \right) dh, \ \delta > 0.$$

Since f_{δ} is a definite Lebesgue integral, it is absolutely continuous on $\mathbb{T}$ and hence

$$|f'_{\delta}(x)| = \frac{2}{\delta} \left| \int_{\delta/2}^{\delta} \frac{1}{h} (f(x+h) - f(x))\, dh \right|.$$

Integrating by parts, we have

$$\int_{\delta/2}^{\delta} \frac{1}{h}(f(x+h)-f(x))\,dh = \left.\frac{1}{h}\int_{\delta/2}^{h}[f(x+t)-f(x)]\,dt\right|_{h=\delta/2}^{h=\delta} + \int_{\delta/2}^{\delta}\frac{dh}{h^2}\int_{\delta/2}^{h}[f(x+t)-f(x)]\,dt$$

$$= \frac{1}{\delta}\int_{\delta/2}^{\delta}[f(x+t)-f(x)]\,dt + \int_{\delta/2}^{\delta}\frac{dh}{h^2}\int_{\delta/2}^{h}[f(x+t)-f(x)]\,dt \tag{3.17}$$

and then

$$\|f'_\delta\|_{p(\cdot),\omega} = \left\|\frac{2}{\delta}\int_{\delta/2}^{\delta}\frac{1}{h}(f(x+h)-f(x))\,dh\right\|_{p(\cdot),\omega}$$

$$\leq \frac{2}{\delta}\left\|\frac{1}{\delta}\int_{\delta/2}^{\delta}[f(x+t)-f(x)]\,dt\right\|_{p(\cdot),\omega} + \frac{2}{\delta}\left\|\int_{\delta/2}^{\delta}\frac{dh}{h^2}\int_{\delta/2}^{h}[f(x+t)-f(x)]\,dt\right\|_{p(\cdot),\omega}. \tag{3.18}$$

For the first term on the right-hand side of (3.18) we have

$$\left\|\frac{1}{\delta}\int_{\delta/2}^{\delta}[f(x+t)-f(x)]\,dt\right\|_{p(\cdot),\omega}$$

$$= \left\|\frac{1}{\delta}\int_{0}^{\delta}[f(x+t)-f(x)]\,dt - \frac{1}{2}\frac{1}{\delta/2}\int_{0}^{\delta/2}[f(x+t)-f(x)]\,dt\right\|_{p(\cdot),\omega}$$

$$\leq \left\|\frac{1}{\delta}\int_{0}^{\delta}[f(x+t)-f(x)]\,dt\right\|_{p(\cdot),\omega} + \frac{1}{2}\left\|\frac{1}{\delta/2}\int_{0}^{\delta/2}[f(x+t)-f(x)]\,dt\right\|_{p(\cdot),\omega}$$

$$\leq \Omega(f,\delta)_{p(\cdot),\omega} + \frac{1}{2}\Omega(f,\delta/2)_{p(\cdot),\omega} \leq \frac{3}{2}\Omega(f,\delta)_{p(\cdot),\omega}. \tag{3.19}$$

For the second term on the right-hand side of (3.18) we get

$$\left\|\int_{\delta/2}^{\delta}\frac{dh}{h^2}\int_{\delta/2}^{h}[f(x+t)-f(x)]\,dt\right\|_{p(\cdot),\omega} =$$

$$= \left\| \int_{\delta/2}^{\delta} \frac{dh}{h^2} \left[\int_0^h [f(x+t) - f(x)]\, dt - \int_0^{\delta/2} [f(x+t) - f(x)]\, dt \right] \right\|_{p(\cdot),\omega}$$

$$= \left\| \int_{\delta/2}^{\delta} \frac{dh}{h} \left[\frac{1}{h} \int_0^h [f(x+t) - f(x)]\, dt - \frac{\delta/2}{h} \frac{1}{\delta/2} \int_0^{\delta/2} [f(x+t) - f(x)]\, dt \right] \right\|_{p(\cdot),\omega}$$

$$\leq c_{15}(p) \int_{\delta/2}^{\delta} \frac{dh}{h} \left(\left\| \frac{1}{h} \int_0^h [f(x+t) - f(x)]\, dt \right\|_{p(\cdot),\omega} + \left\| \frac{2}{\delta} \int_0^{\delta/2} [f(x+t) - f(x)]\, dt \right\|_{p(\cdot),\omega} \right)$$

$$\leq c_{16}(p) \left(\Omega(f,\delta)_{p(\cdot),\omega} + \Omega(f,\delta/2)_{p(\cdot),\omega} \right) \int_{\delta/2}^{\delta} \frac{dh}{h} \leq c_{17}(p) \Omega(f,\delta)_{p(\cdot),\omega}. \tag{3.20}$$

Hence by (3.18), (3.19) and (3.20) we obtain

$$\left\| f'_{\delta} \right\|_{p(\cdot),\omega} \leq c_{18}(p) \delta^{-1} \Omega(f,\delta)_{p(\cdot),\omega}. \tag{3.21}$$

On the other hand, by estimate (3.16)

$$\| f_{\delta}(x) - f(x) \|_{p(\cdot),\omega} = \left\| \frac{2}{\delta} \int_{\delta/2}^{\delta} \left(\frac{1}{h} \int_0^h [f(x+t) - f(x)]\, dt \right) dh \right\|_{p(\cdot),\omega}$$

$$\leq c_{19}(p) \sup_{h \leq \delta} \left\| \frac{1}{h} \int_0^h [f(x+t) - f(x)]\, dt \right\|_{p(\cdot),\omega} = \mathcal{O}\left(\Omega(f,\delta)_{p(\cdot),\omega} \right). \tag{3.22}$$

Now, choosing $\delta := 1/n$, applying Lemma 2.13 and using (3.21) and (3.22) we conclude that

$$E_n(f)_{p(\cdot),\omega} \leq E_n\left(f - f_{1/n}\right)_{p(\cdot),\omega} + E_n\left(f_{1/n}\right)_{p(\cdot),\omega}$$

$$\leq c(p) \left\| f - f_{1/n} \right\|_{p(\cdot),\omega} + \frac{c(p)}{n} \left\| f'_{1/n} \right\|_{p(\cdot),\omega}$$

$$\leq \mathscr{O}\left(\Omega(f,1/n)_{p(\cdot),\omega}\right)+\frac{c(p)}{n}n\Omega(f,1/n)_{p(\cdot),\omega}=\mathscr{O}\left(\Omega(f,1/n)_{p(\cdot),\omega}\right).$$

□

Before the formulation of inverse theorem in $L_{\omega}^{p(\cdot)}(T)$ we will prove some auxiliary results in this space.

Lemma 2.15. *Let $p(\cdot)\in\mathscr{P}_0(T)$, $\omega\in A_{p(\cdot)}(T)$ and T_n be a trigonometric polynomial of degree n. Then there is a positive constant $c(p)$ such that for any $\delta>0$ and $n\in\mathbb{N}$ the inequality*

$$\Omega(T_n,\delta)_{p(\cdot),\omega}\leq c(p)\,\delta\left\|T_n'\right\|_{p(\cdot),\omega}$$

holds.

Proof. Since the maximal operator $Mf: f(x)\rightarrow\sup_{I\ni x}\frac{1}{|I|}\int_I|f(t)|\,dt,\,x\in T$, is a bounded operator in $L_{\omega}^{p(\cdot)}(T)$, $p(\cdot)\in\mathscr{P}_0(T)$, $\omega(\cdot)\in A_{p(\cdot)}(T)$, by Minkowski's inequality we have

$$\begin{aligned}
\Omega(T_n,\delta)_{p(\cdot)} &= \sup_{0<h<\delta}\left\|\frac{1}{h}\int_0^h[T_n(\cdot)-T_n(\cdot+t)]\,dt\right\|_{p(\cdot),\omega}\\
&\leq c_{20}(p)\sup_{0<h<\delta}\frac{1}{h}\int_0^h\|T_n(\cdot)-T_n(\cdot+t)\|_{p(\cdot),\omega}dt\\
&= c_{20}(p)\sup_{0<h<\delta}\frac{1}{h}\int_0^h t\left\|\frac{1}{t}\int_x^{x+t}T_n'(u)\,du\right\|_{p(\cdot),\omega}dt\\
&\leq c_{21}(p)\sup_{0<h<\delta}\frac{1}{h}\int_0^h t\left\|\frac{1}{t}\int_x^{x+t}\left|T_n'(u)\right|du\right\|_{p(\cdot),\omega}dt\\
&\leq c_{22}(p)\sup_{0<h<\delta}\frac{1}{h}\int_0^h t\left\|T_n'\right\|_{p(\cdot),\omega}dt\leq c(p)\,\delta\left\|T_n'\right\|_{p(\cdot),\omega}.
\end{aligned}$$

□

We will use the following simplified version of a general extrapolation theorem proved in [103, Theorem 2.7].

Theorem 2.16. *Suppose that for some* $p_0 > 1$ *the operator* A *is bounded on the space* $L^{p_0}_{\omega}(T)$ *for all weights* $\omega \in A_{p_0}(T)$, *i.e.*

$$\|A(f)\|_{p_0,\omega} \leq c(p_0)\|f\|_{p_0,\omega}, \qquad f \in L^{p_0}_{\omega}([0,2\pi]).$$

If the maximal operator $M(f)$ *is bounded in* $L^{p(\cdot)}_{\omega}(T)$, $p(\cdot) \in \mathscr{P}_0(T)$, $\omega \in A_{p(\cdot)}(T)$ *then there exists a positive constant* $c(p)$ *such that the inequality*

$$\|A(f)\|_{p(\cdot),\omega} \leq c(p_0)\|f\|_{p(\cdot),\omega}$$

holds.

Let $D_n(t)$ and $F_n(t)$ be the Dirichlet and Fejer kernels of order n, respectively, defined by

$$D_n(t) = \frac{1}{2} + \sum_{k=1}^{n} \cos kt \quad \text{and} \quad F_n(t) = \frac{1}{n+1}\sum_{k=0}^{n} D_k(t).$$

We consider the sequence $\{K_n(f)\}$ of the arithmetic means of $S_n(f)$, that is

$$K_n(f)(x) := \frac{1}{n+1}\sum_{k=0}^{n} S_k(f)(x), \quad n = 0,1,2,\ldots.$$

It is known [104, pp. 3] that

$$K_n(f)(x) = \frac{1}{\pi}\int_{\mathbb{T}} f(t)F_n(x-t)\,dt.$$

Since the operator $K_n(f)$ is bounded [144] in the classical weighted Lebesgue spaces $L^{p_0}_{\omega}(T)$, $p_0 > 1$, applying Theorem 2.16 in the case of $A(f) := K_n(f)$ we have

Lemma 2.17. *Let* $p(\cdot) \in \mathscr{P}_0(T)$ *and* $\omega(\cdot) \in A_{p(\cdot)}(T)$. *There exists a positive constant* $c(p)$ *such that for every* $f \in L^{p(\cdot)}_{\omega}(T)$

$$\|K_n(f)\|_{p(\cdot),\omega} \leq c(p)\|f\|_{p(\cdot),\omega}.$$

Lemma 2.18. *Let* $f \in L^{p(\cdot)}_{\omega}(T)$, $p(\cdot) \in \mathscr{P}_0(T)$, $\omega(\cdot) \in A_{p(\cdot)}(T)$. *If* T_n *is a trigonometric polynomial of degree n, then there exists a positive constant* $c(p)$ *such that for* $r = 1,2,3,\ldots$ *the inequality*

$$\left\|T_n^{(r)}\right\|_{p(\cdot),\omega} \leq c(p)n^r\|T_n\|_{p(\cdot),\omega}, \; n = 1,2,\ldots$$

holds.

Proof. We use the technique developed in [163, Vol I, pp. 118]. Since

$$T_n(x) = S_n(T_n)(x) = \frac{1}{\pi}\int_{\mathbb{T}} T_n(u) D_n(u-x)\,du$$

and

$$T_n'(x) = -\frac{1}{\pi}\int_{\mathbb{T}} T_n(u) D_n'(u-x)\,du = \frac{1}{\pi}\int_{\mathbb{T}} T_n(u+x)\left(\sum_{k=1}^{n} k\sin ku\right)du,$$

considering

$$\int_{\mathbb{T}} T_n(u+x)\sum_{k=1}^{n-1} k\sin\left[(2n-k)u\right]du = 0$$

we have

$$T_n'(x) = \frac{1}{\pi}\int_{\mathbb{T}} T_n(u+x)\left(\sum_{k=1}^{n} k\sin ku + \sum_{k=1}^{n-1} k\sin\left[(2n-k)u\right]\right)du$$

$$= \frac{1}{\pi}\int_{\mathbb{T}} T_n(u+x)\,2n\sin nu\left(\frac{1}{2} + \sum_{k=1}^{n-1}\frac{n-k}{n}\cos ku\right)du$$

$$= \frac{2n}{\pi}\int_{\mathbb{T}} T_n(u+x)\sin nu F_{n-1}(u)\,du.$$

Since F_{n-1} is nonnegative,

$$\left|T_n'(x)\right| \leq \frac{2n}{\pi}\int_{\mathbb{T}} \left|T_n(u+x)\right| F_{n-1}(u)\,du$$

$$= \frac{2n}{\pi}\int_{\mathbb{T}} \left|T_n(u)\right| F_{n-1}(u-x)\,du = 2nK_{n-1}(\left|T_n\right|, x)$$

and hence by Lemma 2.17 we get $\left\|T_n'\right\|_{p(\cdot),\omega} \leq c(p)n\left\|T_n\right\|_{p(\cdot),\omega}$. Now by iteration, we obtain the desired relation. □

Let $n \in \mathbb{N}$ and let $m \in \mathbb{N}$ be the number satisfying the inequality $2^{m-1} \leq n < 2^m$. We construct the set of numbers n_j, $j = 1, 2, ..., m-1$ such that $n_0 = 0$, $n_j = 2^j$ and $n_m = n$. It is known [104, p. 209] that

$$2^{(i+1)k} E_{2^i}(f)_{p(\cdot),\omega} \leq 2^{2k} \sum_{\mu=2^{i-1}+1}^{2^i} \mu^{k-1} E_\mu(f)_{p(\cdot),\omega}, \ j \geq 1. \tag{3.23}$$

Now we can formulate the inverse theorem in $L_\omega^{p(\cdot)}(T)$.

Theorem 2.19. *Let $p(\cdot) \in \mathscr{P}_0(T)$ and $\omega(\cdot) \in A_{p(\cdot)}(T)$. Then there exists a positive constant $c(p)$ such that for every $f \in L_\omega^{p(\cdot)}(T)$ the inequality*

$$\Omega(f,1/n)_{p(\cdot),\omega} \leq \frac{c(p)}{n}\sum_{k=0}^{n} E_k(f)_{p(\cdot),\omega} \quad, n=1,2,...$$

holds.

Proof. Let T_n^0 be the best approximation trigonometric polynomial for $f \in L_\omega^{p(\cdot)}(T)$, and let $m \in \mathbb{N}$ be the number such that $2^m \leq n < 2^{m+1}$ for a given $n \in \mathbb{N}$. It is clear that

$$\Omega(f,1/n)_{p(\cdot),\omega} \leq \Omega\left(f-T_{2^{m+1}}^0,1/n\right)_{p(\cdot),\omega} + \Omega\left(T_{2^{m+1}}^0,1/n\right)_{p(\cdot),\omega}. \tag{3.24}$$

Using Minkowsky's inequality and the inequality (3.23) we have

$$\Omega\left(f-T_{2^{m+1}}^0,1/n\right)_{p(\cdot),\omega} \leq c_{23}(p)\left\|f-T_{2^{m+1}}^0\right\|_{p(\cdot),\omega} = c_{23}(p)E_{2^{m+1}}(f)_{p(\cdot),\omega}$$

$$\leq c_{24}(p)\frac{2^{(m+1)}}{n}E_{2^m}(f)_{p(\cdot),\omega} \leq \frac{4c_{24}(p)}{n}\sum_{k=2^{m-1}+1}^{2^m} E_k(f)_{p(\cdot),\omega}. \tag{3.25}$$

On the other hand, applying Lemmas 2.15 and 2.18 and using (3.23) we get

$$\Omega\left(T_{2^{m+1}}^0,1/n\right)_{p(\cdot),\omega} \leq \frac{c(p)}{n}\left\|\left(T_{2^{m+1}}^0\right)'\right\|_{p(\cdot),\omega} = \frac{c(p)}{n}\left\|\left(T_1^0\right)' + \sum_{\nu=0}^{m}\left[\left(T_{2^{\nu+1}}^0\right)' - \left(T_{2^\nu}^0\right)'\right]\right\|_{p(\cdot),\omega}$$

$$\leq \frac{c(p)}{n}\left(\left\|\left(T_1^0\right)'\right\|_{p(\cdot),\omega} + \left\|\sum_{\nu=0}^{m}\left[\left(T_{2^{\nu+1}}^0\right)' - \left(T_{2^\nu}^0\right)'\right]\right\|_{p(\cdot),\omega}\right)$$

$$\leq \frac{c_{25}(p)}{n}\left(\left\|T_1^0\right\|_{p(\cdot),\omega} + \sum_{\nu=0}^{m} 2^{\nu+1}\left\|T_{2^{\nu+1}}^0 - T_{2^\nu}^0\right\|_{p(\cdot),\omega}\right)$$

$$\leq \frac{c_{26}(p)}{n}\left(E_0(f)_{p(\cdot),\omega} + \sum_{\nu=0}^{m} 2^{\nu+1}E_{2^\nu}(f)_{p(\cdot),\omega}\right)$$

$$= \frac{c_{26}(p)}{n}\left(E_0(f)_{p(\cdot),\omega} + 2E_1(f)_{p(\cdot),\omega} + \sum_{\nu=1}^{m} 2^{\nu+1}E_{2^\nu}(f)_{p(\cdot),\omega}\right)$$

$$\leq \frac{c_{27}(p)}{n}\left(3E_0(f)_{p(\cdot),\omega} + \sum_{\nu=1}^{m}\sum_{k=2^{\nu-1}+1}^{2^\nu} E_k(f)_{p(\cdot),\omega}\right)$$

$$\leq \frac{c_{28}(p)}{n}\left(E_0(f)_{p(\cdot),\omega} + \sum_{k=1}^{2^m} E_k(f)_{p(\cdot),\omega}\right) = \frac{c_{28}(p)}{n}\sum_{k=0}^{2^m} E_k(f)_{p(\cdot),\omega}. \tag{3.26}$$

Now combining (3.24), (3.25) and (3.26) we conclude that

$$\Omega(f,1/n)_{p(\cdot),\omega} \leq \frac{c_{29}(p)}{n}\left(\sum_{k=2^{m-1}+1}^{2^m} E_k(f)_{p(\cdot),\omega} + \sum_{k=0}^{2^m} E_k(f)_{p(\cdot),\omega}\right)$$
$$\leq \frac{c(p)}{n}\sum_{k=0}^{2^m} E_k(f)_{p(\cdot),\omega}.$$

□

We also can give the constructive characterization of variable exponent weighted Lipschitz class $Lip(\alpha, p(\cdot), \omega)$, $0<\alpha<1$.

Corollary 2.20. *Let* $E_n(f)_{p(\cdot),\omega} = \mathscr{O}(n^{-\alpha})$, $\omega \in A_{p(\cdot)}(T)$, $p(\cdot) \in \mathscr{P}_0(T)$, $\alpha > 0$, *then under the conditions of Theorem 2.19 we have*

$$\Omega(f,\delta)_{p(\cdot),\omega} = \begin{cases} \mathscr{O}(\delta^\alpha) & ,\alpha<1 \\ \mathscr{O}(\delta^\alpha \log(1/\delta)) & ,\alpha=1 \\ \mathscr{O}(\delta^r) & ,1<\alpha. \end{cases}$$

Hence, if we define a generalized Lipschitz class $Lip^{p(\cdot)}_{\alpha,\omega}(T)$ as

$$Lip^{p(\cdot)}_{\alpha,r,\omega}(T) := \left\{ f \in L^{p(\cdot)}(T) : \Omega(f,\delta)_{p(\cdot),\omega} = \mathscr{O}(\delta^\alpha),\ \ \delta>0 \right\},$$

then in particular we have

Corollary 2.21. *Let* $E_n(f)_{p(\cdot),\omega} = \mathscr{O}(n^{-\alpha})$, $\omega \in A_{p(\cdot)}(T)$, $p(\cdot) \in \mathscr{P}_0(T)$. *If* $E_n(f)_{p(\cdot),\omega} = \mathscr{O}(n^{-\alpha})$ *for* $\alpha \in (0,1)$, *then* $f \in Lip^{p(\cdot)}_{\alpha,\omega}(T)$.

On the other hand, from Theorem 2.14 we get

Corollary 2.22. *If* $f \in Lip^{p(\cdot)}_{\alpha,\omega}(T)$ *with* $p(\cdot) \in \mathscr{P}_0(T)$ *and some* $\alpha > 0$, *then* $E_n(f)_{p(\cdot)} = \mathscr{O}(n^{-\alpha})$.

Now Corollaries 2.21 and 2.22 give the following constructive characterization of $Lip^{p(\cdot)}_{\alpha,\omega}(T)$:

Theorem 2.23. *Let* $f \in Lip^{p(\cdot)}_{\alpha,\omega}(T)$, $p(\cdot) \in \mathscr{P}_0(T)$ *and* $\alpha > 0$. *The following statements are equivalent:*

$$i)\ f \in Lip^{p(\cdot)}_{\alpha,\omega}(T),$$
$$ii)\ E_n(f)_{p(\cdot),\omega} = \mathscr{O}(n^{-\alpha}), n \in \mathbb{N}.$$

Note that for higher orders of moduli of smoothness the results obtained in this subsection were proved in [156].

3. Approximation in variable exponent Smirnov classes of functions defined on domains of complex plane

In this section, we define the variable exponent Smirnov classes as a generalization of the classical Smirnov classes of analytic functions and study approximation problems in these classes. More detailed information on the main results presented in this section can be found in [120]- [122] and [124].

3.1. Definitions

Let G be a bounded simple connected domain on the complex plane $\mathbb{C}$ bounded by a rectifiable Jordan curve Γ and let $G^{-} := ext\ \Gamma$. Let also $\mathbb{U} := \{w \in \mathbb{C} : |w| < 1\}$, $\mathbb{T} := \partial\mathbb{U}$ and $\mathbb{U}^{-} := \overline{\mathbb{C}} \setminus \overline{\mathbb{U}}$. By $L^{p}(\Gamma)$, $1 \leq p < \infty$, we denote the set of all measurable complex valued functions f on Γ such that $|f|^{p}$ is Lebesgue integrable with respect to arc length on Γ. We recall that if for a given analytic function f on G there exists a sequence of rectifiable Jordan curves (γ_n) in G tending to the boundary Γ in the sense that γ_n eventually surrounds each compact subdomain of G such that

$$\int_{\gamma_n} |f(z)|^{p}\,|dz| \leq C < \infty,$$

then we say that f belongs to the Smirnov class $E^{p}(G)$. Each function $f \in E^{p}(G)$ has a non-tangential limit almost everywhere $(a.e.)$ on Γ and the boundary function belongs to $L^{p}(\Gamma)$ [109, p. 419-438].The Smirnov space $E^{p}\left(G^{-}\right)$ is defined similarly. Throughout this work we suppose that $f(\infty) = 0$ as soon as $f \in E^{p(\cdot)}(G^{-})$.

Let $\mathscr{E}$ be the segment $[0, 2\pi]$ or a Jordan rectifiable curve Γ and let $p(\cdot) : \mathscr{E} \to \mathbb{R}^{+} := [0, \infty)$ be a Lebesgue measurable function defined on $\mathscr{E}$ such that

$$1 \leq p_{-} := ess \inf_{z\in\mathscr{E}} p(z) \leq ess \sup_{z\in\mathscr{E}} p(z) =: p^{+} < \infty. \tag{3.27}$$

Definition 3.1. *We say that $p(\cdot) \in \mathscr{P}(\mathscr{E})$, if $p(\cdot)$ satisfies the conditions* (3.27) *and the inequality*

$$|p(z_1) - p(z_2)| \ln\left(\frac{|\mathscr{E}|}{|z_1 - z_2|}\right) \leq c, \quad \forall z_1, z_2 \in \mathscr{E},$$

with a positive constant $c(p)$, where $|\mathscr{E}|$ is the Lebesgue measure of $\mathscr{E}$. If $p(\cdot) \in \mathscr{P}(\mathscr{E})$ and $p_{-} > 1$, then we say that $p(\cdot) \in \mathscr{P}_0(\mathscr{E})$.

Definition 3.2. *We define the variable exponent Lebesgue space* $L^{p(\cdot)}(\Gamma)$, $p(\cdot) \in \mathscr{P}_0(\mathscr{E})$, *as a set of Lebesgue measurable functions* f *such that* $\int_\Gamma |f(z)|^{p(z)} |dz| \leq C < \infty$.

Equipped with the norm

$$\|f\|_{L^{p(\cdot)}(\Gamma)} := \inf\left\{\lambda > 0 : \int_\Gamma |f(z)/\lambda|^{p(z)} |dz| \leq 1\right\} < \infty,$$

$L^{p(\cdot)}(\Gamma)$ becomes a Banach space. In the case of $\Gamma := \mathbb{T}$, we obtain the variable exponent Lebesgue space $L^{p(\cdot)}(\mathbb{T})$ with the norm

$$\|f\|_{L^{p(\cdot)}(\mathbb{T})} := \inf\left\{\lambda > 0 : \int_0^{2\pi} \left|f\left(e^{it}\right)/\lambda\right|^{p\left(e^{it}\right)} |dt| \leq 1\right\} =: \|f\|_{L^{p(\cdot)}([0,2\pi])}.$$

Definition 3.3. *Let* $p(\cdot) \in \mathscr{P}_0(\mathscr{E})$. *The sets*

$$E^{p(\cdot)}(G) := \left\{f \in E^1(G) : f \in L^{p(\cdot)}(\Gamma)\right\},$$

$$E^{p(\cdot)}(G^-) := \left\{f \in E^1(G^-) : f \in L^{p(\cdot)}(\Gamma)\right\}$$

are called the variable exponent Smirnov classes of analytic functions in G *and* G^-, *respectively.*

Equipping them with the norms

$$\|f\|_{E^{p(\cdot)}(G)} := \|f\|_{L^{p(\cdot)}(\Gamma)}, \quad \|f\|_{E^{p(\cdot)}(G^-)} := \|f\|_{L^{p(\cdot)}(\Gamma)}$$

we make $E^{p(\cdot)}(G)$ and $E^p\left(G^-\right)$ the Banach spaces.

Let g be a continuous function and let

$$\omega(g,t) := \sup_{|t_1 - t_2| \leq t} |g(t_1) - g(t_2)|, \quad t > 0$$

be its modulus of continuity.

Definition 3.4. *Let* Γ *be a smooth Jordan curve and let* $\theta(s)$ *be the angle between the tangent and the positive real axis expressed as a function of arc length* s. *If* Γ *has a modulus of continuity* $\omega(\theta, s)$, *satisfying the Dini-smooth condition*

$$\int_0^\delta [\omega(\theta,s)/s]\, ds < \infty, \; \delta > 0,$$

then we say that Γ *is a Dini-smooth curve.*

We denote the set of Dini-smooth curves by $\mathfrak{D}$.

By φ and φ_1 we denote the conformal mappings of G^- and G onto $\mathbb{U}^-$, respectively, normalized by the conditions

$$\varphi(\infty)=\infty,\ \lim_{z\longrightarrow\infty}\varphi(z)/z>0 \quad \text{and} \quad \varphi_1(0)=0,\ \lim_{z\to 0}z\varphi_1(z)>0.$$

Let ψ and ψ_1 be the inverse mappings of φ and φ_1, respectively. The pairs (φ,φ_1) and (ψ,ψ_1) have continuous extensions to Γ and $\mathbb{T}$, respectively. Their derivatives (φ',φ_1') and (ψ',ψ_1') have definite nontangential boundary values a.e. on Γ and $\mathbb{T}$, and the boundary functions are integrable with respect to Lebesgue measure on Γ and $\mathbb{T}$, respectively [109, p. 419-438].

If $\Gamma\in\mathfrak{D}$, then by [161] there are positive constants $c_i>0, i=1,2,..,8$ such that

$$\begin{aligned} 0 &< c_1\le\left|\psi'(w)\right|\le c_2<\infty,\ 0<c_3\le\left|\varphi'(z)\right|\le c_4<\infty, \\ 0 &< c_5\le\left|\psi_1'(w)\right|\le c_6<\infty,\ 0<c_7\le\left|\varphi_1'(z)\right|\le c_8<\infty, \end{aligned} \tag{3.28}$$

a.e. on $\mathbb{T}$ and Γ, respectively. For $f\in L^{p(\cdot)}(\Gamma)$, $p\in\mathscr{P}(\Gamma)$, we set $f_0:=f\circ\psi$, $p_0:=p\circ\psi$, $f_1:=f\circ\psi_1$ and $p_1:=p\circ\psi_1$. If $\Gamma\in\mathfrak{D}$, then using (3.28) we can show that the following relations hold:

$$\begin{aligned} f_1 &\in L^{p_1(\cdot)}(\mathbb{T})\Leftrightarrow f\in L^{p(\cdot)}(\Gamma)\Leftrightarrow f_0\in L^{p_0(\cdot)}(\mathbb{T}) \\ p_0 &\in \mathscr{P}(\mathbb{T})\Leftrightarrow p\in\mathscr{P}(\Gamma)\Leftrightarrow p_1\in\mathscr{P}(\mathbb{T}). \end{aligned}$$

For a given function $f\in L^{p(\cdot)}(\Gamma)$ we define the Cauchy type integrals

$$f_0^+(w):=\frac{1}{2\pi i}\int_{\mathbb{T}}\frac{f_0(\tau)}{\tau-w}d\tau,\ w\in\mathbb{U} \quad\text{and}\quad f_1^+(w):=\frac{1}{2\pi i}\int_{\mathbb{T}}\frac{f_1(\tau)}{\tau-w}d\tau,\ \ w\in\mathbb{U},$$

which are analytic in $\mathbb{U}$.

Definition 3.5. *For $f\in L^{p(\cdot)}(\mathbb{T})$, $p(\cdot)\in\mathscr{P}(\mathbb{T})$ and $t>0$ we set*

$$\Delta_t^r f(w):=\sum_{s=0}^{r}(-1)^{r+s}\binom{r}{s}f\left(we^{ist}\right), \quad r=1,2,3,...$$

and define the rth modulus of smoothness by

$$\Omega_r(f,\delta)_{\mathbb{T},p(\cdot)}:=\sup_{0<|h|\le\delta}\left\|\frac{1}{h}\int_0^h\Delta_t^r f(w)\,dt\right\|_{L^{p(\cdot)}(\mathbb{T})}.$$

We also define the modulus of smoothness for $f\in E^{p(\cdot)}(G)$ *and* $f\in E^{p(\cdot)}(G^{-})$ *as follows:*

$$\begin{aligned}\Omega_r(f,\delta)_{G,p(\cdot)} &:= \Omega\left(f_0^{+},\delta\right)_{\mathbb{T},p_0(\cdot)},\\ \Omega_r(f,\delta)_{G^{-},p(\cdot)} &:= \Omega\left(f_1^{+},\delta\right)_{\mathbb{T},p_1(\cdot)}, \quad \delta>0.\end{aligned}$$

For construction of the approximation aggregates, we use the Faber polynomials $F_k(z)$ and $\widetilde{F}_k(1/z)$ with respect to z and $1/z$, respectively, defined by the representations [100] (see, also [154]):

$$\frac{\psi'(w)}{\psi(w)-z}=\sum_{k=0}^{\infty}\frac{F_k(z)}{w^{k+1}}, \quad w\in\mathbb{U}^{-}, \quad z\in G, \tag{3.29}$$

$$\frac{\psi_1'(w)}{\psi_1(w)-z}=\sum_{k=1}^{\infty}-\frac{\widetilde{F}_k(1/z)}{w^{k+1}}, \quad w\in\mathbb{U}^{-}, \; z\in G^{-}. \tag{3.30}$$

Using (3.29) and Cauchy's integral formula

$$f(z)=\int_{\Gamma}\frac{f(\zeta)}{\zeta-z}d\zeta=\frac{1}{2\pi i}\int_{|w|=1}\frac{f_0(w)\psi'(w)}{\psi(w)-z}dw, \; z\in G,$$

which holds for every $f\in E^{p(\cdot)}(G)\subset E^{1}(G)$, we have

$$f(z)\sim\sum_{k=0}^{\infty}a_kF_k(z)\;, z\in G, \tag{3.31}$$

where

$$a_k=a_k(f):=\frac{1}{2\pi i}\int_{\mathbb{T}}\frac{f_0(w)}{w^{k+1}}dw, \quad k=0,1,2,... \tag{3.32}$$

Similarly, using (3.30) and the integral representation

$$f(z)=\int_{\Gamma}\frac{f(\zeta)}{\zeta-z}d\zeta=\frac{1}{2\pi i}\int_{|w|=1}\frac{f_1(w)\psi_1'(w)}{\psi_1(w)-z}dw, \; z\in G,$$

which holds for every $f\in E^{p(\cdot)}(G^{-})\subset E^{1}(G^{-})$, we have

$$f(z)\sim\sum_{k=0}^{\infty}\widetilde{a}_k\widetilde{F}_k(1/z), \; z\in G^{-} \tag{3.33}$$

with the coefficients

$$\widetilde{a}_k=\widetilde{a}_k(f):=\frac{1}{2\pi i}\int_{\mathbb{T}}\frac{f_1(w)}{w^{k+1}}dw, \quad k=1,2,\dots. \tag{3.34}$$

3.2. Auxiliary results

Let $f \in L^{p(\cdot)}(\Gamma)$, $p(\cdot) \in \mathscr{P}_0(\Gamma)$, and let

$$S_\Gamma(f)(z) := \lim_{\varepsilon \to 0} \frac{1}{2\pi i} \int\limits_{\Gamma \setminus \{\zeta \in \Gamma : |\zeta - z| < \varepsilon\}} \frac{f(\zeta)}{\zeta - z} d\zeta$$

be its Cauchy singular integral. By Privalov's theorem, the Cauchy type integrals

$$f^+(z) := \frac{1}{2\pi i} \int\limits_\Gamma \frac{f(\zeta)}{\zeta - z} d\zeta = \frac{1}{2\pi i} \int\limits_{\mathbb{T}} \frac{[\psi'(w)]}{\psi(w) - z} f_0(w)\, dw, \quad z \in G,$$

$$f^-(z) := \frac{1}{2\pi i} \int\limits_\Gamma \frac{f(\zeta)}{\zeta - z} d\zeta = \frac{1}{2\pi i} \int\limits_{\mathbb{T}} \frac{[\psi'(w)]}{\psi(w) - z} f_1(w)\, dw, \quad z \in G^-,$$

have the nontangential inside and outside limits f^+ and f^-, respectively, *a.e.* on Γ. Furthermore, the formulas

$$f^+(z) = S_\Gamma(f)(z) + \frac{1}{2} f(z) \quad \text{and} \quad f^-(z) = S_\Gamma(f)(z) - \frac{1}{2} f(z) \tag{3.35}$$

are valid a.e. on Γ, which imply that

$$f(z) = f^+(z) - f^-(z) \tag{3.36}$$

a.e. on Γ.

The following theorem is a special case of the more general result on the boundedness of Cauchy singular operator $S_\Gamma(f)$ in $L^{p(\cdot)}(\Gamma)$, proved in [140, Theorem 4.21].

Theorem 3.1. *Let $\Gamma \in \mathfrak{D}$ and $p \in \mathscr{P}_0(\Gamma)$. Then the Cauchy singular operator $S_\Gamma(f)$ is bounded in $L^{p(\cdot)}(\Gamma)$.*

Lemma 3.2 ([121]). *Let $\Gamma \in \mathfrak{D}$. If $f \in L^{p(\cdot)}(\Gamma)$, $p \in \mathscr{P}_0(\Gamma)$, then $f^+ \in E^{p(\cdot)}(G)$ and $f^- \in E^{p(\cdot)}(G^-)$.*

Now we consider the operators

$$T(f)(z) := \frac{1}{2\pi i} \int\limits_{\mathbb{T}} \frac{f(w)\, \psi'(w)}{\psi(w) - z} dw, \quad f \in E^{p_0(\cdot)}(\mathbb{U}), \ z \in G,$$

$$\widetilde{T}(f)(z) := \frac{1}{2\pi i} \int\limits_{\mathbb{T}} \frac{f(w)\, \psi_1'(w)}{\psi_1(w) - z} dw, \quad f \in E^{p_1(\cdot)}(\mathbb{U}), \ z \in G^-,$$

defined on the classes $E^{p_0(\cdot)}(\mathbb{U})$ and $E^{p_1(\cdot)}(\mathbb{U})$, respectively.

The following theorem plays an important role in extending the direct and inverse theorems proved in Subsection 2.2. to the case of $E^{p(\cdot)}(G)$.

Theorem 3.3. *Let* $\Gamma \in \mathfrak{D}$ *and* $p(\cdot) \in \mathscr{P}_0(\Gamma)$. *Then,*

i) the operator $T : E^{p_0(\cdot)}(\mathbb{U}) \to E^{p(\cdot)}(G)$ *is linear, bounded, one-to-one and onto. Moreover,* $T\left(f_0^+\right) = f$ *for every* $f \in E^{p(\cdot)}(G)$

ii) the operator $\widetilde{T} : E^{p_1(\cdot)}(\mathbb{U}) \to E^{p(\cdot)}\left(G^-\right)$ *is linear, one-to-one and onto. If* $f \in E^{p(\cdot)}(G^-)$, *then* $\widetilde{T}\left(f_1^+\right) = f$.

Note that the assertion i) was proved in [121]. The assertion ii) can be proved in a similar way.

Let Π_n be the class of algebraic polynomials of degree not exceeding n and let

$$E_n(f)_{G,p(\cdot)} := \inf\left\{\|f - P_n\|_{L^{p(\cdot)}(\Gamma)} : \ P_n \in \Pi_n\right\}, \ n = 1, 2, ...$$

be the best approximation number of $f \in E^{p(\cdot)}(G)$ in Π_n.

Lemma 3.4. *Let* $\Gamma \in \mathfrak{D}$ *and* $p(\cdot) \in \mathscr{P}_0(\Gamma)$. *Then there exist the positive constants* $c_i(p)$, *i=30,31,32,33, such that the following assertions hold:*

$$i) \text{If } f \in E^{p(\cdot)}(G), \text{ then } E_n\left(f_0^+\right)_{p_0(\cdot)} \le c_{30}(p) E_n(f)_{G,p(\cdot)} \le c_{31}(p) E_n\left(f_0^+\right)_{p_0(\cdot)},$$

$$ii) \text{If } f \in E^{p(\cdot)}\left(G^-\right), \text{ then } E_n\left(f_1^+\right)_{p_1(\cdot)} \le c_{32}(p) E_n(f)_{G^-,p(\cdot)} \le c_{33}(p) E_n\left(f_1^+\right)_{p_1(\cdot)}.$$

Proof. Assertion *i*) was proved in [121]. We will prove *ii*). Since the operator $\widetilde{T} : E^{p_1(\cdot)}(\mathbb{U}) \to E^{p(\cdot)}\left(G^-\right)$ is linear, one-to-one and onto, it has the inverse operator $\widetilde{T}^{-1} : E^{p(\cdot)}\left(G^-\right) \to E^{p_1(\cdot)}(\mathbb{U})$, which is also linear, bounded, one-to-one and onto. If $f \in E^{p(\cdot)}\left(G^-\right)$, then $\widetilde{T}^{-1}(f) = f_1^+ \in E^{p_1(\cdot)}(\mathbb{U})$. If P_n^* is the polynomial of the best approximation to f in $E^{p(\cdot)}\left(G^-\right)$ with respect to $1/z$ and of degree not exceeding n, then $\widetilde{T}^{-1}(P_n^*)$ is a polynomial with respect to w and of degree not exceeding n. Therefore, denoting $c_{34}(p) := \left\|\widetilde{T}^{-1}\right\|$, we get

$$\begin{aligned} E_n\left(f_1^+\right)_{p_1(\cdot)} &\le \left\|f_1^+ - \widetilde{T}^{-1}(P_n^*)\right\|_{L^{p_1(\cdot)}(\mathbb{T})} \\ &\le \left\|\widetilde{T}^{-1}(f) - \widetilde{T}^{-1}(P_n^*)\right\|_{L^{p_1(\cdot)}(\mathbb{T})} \\ &\le \left\|\widetilde{T}^{-1}\right\| \|f - P_n^*\|_{L^{p(\cdot)}(\Gamma)} = c_{34}(p) E_n(f)_{G^-,p(\cdot)}. \end{aligned}$$

On the other hand, by Theorem 3.3 we have $\widetilde{T}\left(f_1^+\right) = f \in E^{p(\cdot)}(G^-)$, and denoting $c_{35}(p) := \left\|\widetilde{T}\right\|$, by boundedness of operator $\widetilde{T}$ we have

$$\begin{aligned} E_n(f)_{G^-,p(\cdot)} &\leq \left\|f - \widetilde{T}\left(P_n^*\right)\right\|_{L^{p(\cdot)}(\Gamma)} \\ &\leq \left\|\widetilde{T}\left(f_1^+\right) - \widetilde{T}\left(P_n^*\right)\right\|_{L^{p(\cdot)}(\Gamma)} \\ &\leq \left\|\widetilde{T}\right\| \left\|f_1^+ - P_n^*\right\|_{L^{p_1(\cdot)}(\mathbb{T})} = c_{35}(p) E_n\left(f_1^+\right)_{p_1(\cdot)}. \end{aligned}$$

□

3.3. Direct and Inverse Theorems in Variable Exponent Smirnov Classes

In this subsection, in terms of the higher order modulus of smoothness, the direct and inverse theorems of approximation theory in the variable exponent Smirnov classes $E^{p(\cdot)}(G)$ are proved. As a corollary, some results on constructive characterization problems in the generalized Lipschitz classes are presented.

The direct and inverse theorems in $E^{p(\cdot)}(G)$ can be formulated as follows:

Theorem 3.5. *Let $\Gamma \in \mathfrak{D}$. If $f \in E^{p(\cdot)}(G)$, $p(\cdot) \in \mathscr{P}_0(\Gamma)$, then there is a positive constant $c(p,r)$ such that the inequality*

$$E_n(f)_{G,p(\cdot)} \leq c(p,r)\,\Omega_r(f,1/n)_{G,p(\cdot)}, \quad n = 1,2,3,\ldots$$

holds.

Theorem 3.6. *Let $\Gamma \in \mathfrak{D}$. If $f \in E^{p(\cdot)}(G)$, $p(\cdot) \in \mathscr{P}_0(\Gamma)$, then there is a positive constant $c(p,r)$ such that the inequality*

$$\Omega_r(f,1/n)_{G,p(\cdot)} \leq \frac{c(p,r)}{n^r} \sum_{k=0}^{n} (k+1)^{r-1} E_k(f)_{G,p(\cdot)}, \quad n = 1,2,3,\ldots$$

holds.

Theorems 3.5 and 3.6 in the case of $r = 1$ were proved in [121] (see also, [122]).

Note that in classical Smirnov classes the direct and inverse problems of approximation theory were investigated adequately. Detailed information about these investigations can be found in [160], [94], [138], [113], [95], [114], [100], [115] and references given therein.

Corollary 3.7. *If $E_n(f)_{G,p(\cdot)} = \mathcal{O}\left(n^{-\alpha}\right)$, $\alpha > 0$, then under the conditions of Theorem 3.6*

$$\Omega_r(f,\delta)_{G,p(\cdot)} = \begin{cases} \mathcal{O}(\delta^{\alpha}), & r > \alpha \\ \mathcal{O}(\delta^{r}\log 1/\delta), & r = \alpha \\ \mathcal{O}(\delta^{r}), & r < \alpha. \end{cases}$$

If we define the generalized Lipschitz class $Lip(G,p(\cdot),\alpha)$ with $\alpha > 0$ and $r := [\alpha]+1$, where $[\alpha]$ is the integer part of α, by

$$Lip(G,p(\cdot),\alpha) := \left\{ f \in E^{p(\cdot)}(G) : \Omega_r(f,\delta)_{G,p(\cdot)} = \mathcal{O}(\delta^{\alpha}),\ \delta > 0 \right\},$$

then from Corollary 3.7 we obtain

Corollary 3.8. *If $E_n(f)_{G,p(\cdot)} = \mathcal{O}\left(n^{-\alpha}\right)$, $\alpha > 0$, then under the conditions of Theorem 3.6 we have $f \in Lip(G,p(\cdot),\alpha)$.*

On the other hand, from Theorem 3.5 we have

Corollary 3.9. *If $f \in Lip(G,p(\cdot),\alpha)$, $\alpha > 0$, then $E_n(f)_{G,p(\cdot)} = \mathcal{O}\left(n^{-\alpha}\right)$.*

Combining Corollaries 3.8 and 3.9 we obtain the constructive characterization of the generalized Lipschitz class $Lip(G,p(\cdot),\alpha)$:

Theorem 3.10. *Let $\Gamma \in \mathfrak{D}$ and $p(\cdot) \in \mathcal{P}_0(\Gamma)$. Then for $\alpha > 0$ the following statements are equivalent:*

$$i) f \in Lip(G,p(\cdot),\alpha), \quad ii) E_n(f)_{G,p(\cdot)} = \mathcal{O}\left(n^{-\alpha}\right) \qquad n = 1,2,3,\ldots.$$

All of the results formulated above can be stated also for the classes $E^{p(\cdot)}\left(G^{-}\right)$.

Let

$$E_n(f)_{G^{-},p(\cdot)} := \inf\left\{ \|f - P_n^*\|_{L^{p(\cdot)}(\Gamma)} : P_n^* \in \Pi_n^* \right\},\ n = 1,2,\ldots$$

be the best approximation number of $f \in E^{p(\cdot)}(G^{-})$ in the class Π_n^* of algebraic polynomials with respect to $1/z$ of degree not exceeding n.

Theorem 3.11. *Let $\Gamma \in \mathfrak{D}$. If $f \in E^{p(\cdot)}\left(G^{-}\right)$, $p(\cdot) \in \mathcal{P}_0(\Gamma)$, then there exists a positive constant $c(p,r)$ such that the inequality*

$$E_n(f)_{G^{-},p(\cdot)} \le c(p,r)\,\Omega_r(f,1/n)_{G^{-},p(\cdot)},\ n = 1,2,3,\ldots$$

holds.

Theorem 3.12. *Let $\Gamma \in \mathfrak{D}$. If $f \in E^{p(\cdot)}(G^-)$ with $p(\cdot) \in \mathscr{P}_0(\Gamma)$, then there exists a positive constant $c(p,r)$ such that the inequality*

$$\Omega_r(f,1/n)_{G^-,p(\cdot)} \leq \frac{c(p,r)}{n^r}\sum_{k=0}^{n}(k+1)^{r-1}E_k(f)_{G^-,p(\cdot)}, \; n=1,2,3,....$$

holds.

Corollary 3.13. *If $E_n(f)_{G^-,p(\cdot)} = \mathscr{O}(n^{-\alpha})$ for some $\alpha > 0$, then under the conditions of Theorem 3.12*

$$\Omega_r(f,\delta)_{G^-,p(\cdot)} = \begin{cases} \mathscr{O}(\delta^\alpha), & r>\alpha \\ \mathscr{O}(\delta^r \log(1/\delta)), & r=\alpha \\ \mathscr{O}(\delta^r), & r<\alpha. \end{cases}$$

Similarly, defining the generalized Lipschitz class $Lip(G^-, p(\cdot), \alpha)$ with $\alpha > 0$ and $r := [\alpha]+1$ by

$$Lip(G^-,p(\cdot),\alpha) := \left\{f \in E^{p(\cdot)}(G^-) : \Omega_r(f,\delta)_{G^-,p(\cdot)} = \mathscr{O}(\delta^\alpha), \; \delta > 0\right\},$$

we have

Corollary 3.14. *If $E_n(f)_{G^-,p(\cdot)} = \mathscr{O}(n^{-\alpha})$, $\alpha > 0$, then under the conditions of Theorem 3.12 we have $f \in Lip(G^-,p(\cdot),\alpha)$.*

On the other hand Theorem 3.11 implies

Corollary 3.15. *If $f \in Lip(G^-,p(\cdot),\alpha)$ for some $\alpha > 0$, then $E_n(f)_{G^-,p(\cdot)} = \mathscr{O}(n^{-\alpha})$.*

Hence by means of Corollaries 3.14 and 3.15 we can formulate the following theorem which gives a constructive characterization of the generalized Lipschitz class $Lip(G^-,p(\cdot),\alpha)$:

Theorem 3.16. *Let $\Gamma \in \mathfrak{D}$, $p(\cdot) \in \mathscr{P}_0(\Gamma)$ and $\alpha > 0$. The following statements are equivalent:*

$$i)\; f \in Lip(G^-,p(\cdot),\alpha), \quad ii)\; E_n(f)_{G^-,p(\cdot)} = \mathscr{O}(n^{-\alpha}) \qquad n=1,2,3,\ldots.$$

3.4. Proofs of Direct and Inverse Theorems

Proof. (**Theorem 3.5**) If $f \in E^{p(\cdot)}(G)$, then $f_0^+ = T^{-1}(f) \in E^{p_0(\cdot)}(\mathbb{U})$. Hence applying second inequality of assertion $i)$ at Lemma 3.4 and the appropriate direct theorem of Section 2. we have

$$\begin{aligned} E_n(f)_{G,p(\cdot)} &\leq c_{31}(p) E_n\left(f_0^+\right)_{p_0(\cdot)} \\ &\leq c_{31}(p) c(p,r) \Omega_r\left(f_0^+, 1/n\right)_{\mathbb{T},p_0(\cdot)} \\ &= c(p,r) \Omega_r(f, 1/n)_{G,p(\cdot)}. \end{aligned}$$

□

Proof. (**Theorem 3.6**) If $f \in E^{p(\cdot)}(G)$, then $f_0^+ \in E^{p_0(\cdot)}(\mathbb{U})$. Applying the appropriate inverse theorem of Section 2. for the boundary values of f_0^+ and first inequality of the step $i)$ at Lemma 3.4 we obtain the desired inequality:

$$\begin{aligned} \Omega_r(f, 1/n)_{G,p(\cdot)} &= \Omega_r\left(f_0^+, 1/n\right)_{\mathbb{T},p_0(\cdot)} \leq \frac{c(p,r)}{n^r} \sum_{k=0}^{n} (k+1)^{r-1} E_k\left(f_0^+\right)_{p_0(\cdot)} \\ &\leq \frac{c(p,r) c_{30}(p)}{n^r} \sum_{k=0}^{n} (k+1)^{r-1} E_n(f)_{G,p(\cdot)} \\ &\leq \frac{c(p,r)}{n^r} \sum_{k=0}^{n} (k+1)^{r-1} E_n(f)_{G,p(\cdot)}. \end{aligned}$$

□

The proofs of Theorems 3.11 and 3.12 go in a similar way, applying the assertion *ii*) of Theorem 3.3 and the assertion ii) of Lemma 3.4.

4. Approximation by Matrix Transforms in Weighted Lebesgue Spaces with Variable Exponent

In this section, we give some results on the approximation properties of matrix transforms in variable exponent spaces. Detailed proofs of these results can be found in [123]. The weighted case was studied in [130].

Let $f \in L^1(T)$ and let

$$f(x) \sim \frac{a_0}{2} + \sum_{k=1}^{\infty} (a_k \cos kx + b_k \sin kx)$$

be its Fourier series, where $a_k := a_k(f)$ and $b_k := b_k(f)$ are the Fourier coefficients of f defined by

$$a_k(f) := \frac{1}{2\pi} \int_0^{2\pi} f(t) \cos(kt)\, dt \quad \text{and} \quad b_k(f) := \frac{1}{2\pi} \int_0^{2\pi} f(t) \sin(kt)\, dt.$$

Let also

$$S_n(f)(x) = \frac{a_0}{2} + \sum_{k=1}^{n} (a_k \cos kx + b_k \sin kx) \ n = 1, 2, ...$$

be the nth partial sums of Fourier series of f.

Let $A = (a_{n,k})$ be an infinite lower triangular regular matrix with non-negative entries and let $s_n^{(A)}$ $(n = 0, 1, ...)$ denote the row sums of this matrix, that is

$$s_n^{(A)} = \sum_{k=0}^{n} a_{n,k}.$$

For a given matrix $A = (a_{n,k})$, the matrix transform of Fourier series of f is defined by

$$T_n^{(A)}(f)(x) := \sum_{k=0}^{n} a_{n,k} S_k(f)(x).$$

If we consider the lower triangular matrix A with entries $a_{n,k} = p_{n-k}/P_n$, then the matrix transform $T_n^{(A)}(f)$ coincides with Nörlund mean $N_n(f)$ defined below.

Let (p_n) be a sequence of positive numbers. The Nörlund means of the Fourier series of f with respect to the sequence (p_n) are defined as

$$N_n(f)(x) := \frac{1}{P_n} \sum_{k=0}^{n} p_{n-k} S_k(f)(x),$$

where $P_n = \sum_{k=0}^{n} p_k$. Clearly, if $p_n = 1$ for all $n = 0, 1, 2, ..,$ then $N_n(f)$ coincides with the Cesàro mean $\sigma_n(f)$, defined as

$$\sigma_n(f)(x) := \frac{1}{n+1} \sum_{k=0}^{n} S_k(f)(x).$$

We will investigate the approximation properties of the matrix transforms $T_n^{(A)}(f)$ in the weighted variable exponent Lebesgue spaces $L_\omega^{p(\cdot)}(T)$, $\omega \in A_{p(\cdot)}(T)$. This problem in classical nonweighted Lebesgue spaces was investigated in [147], [143], [101], [141]; the weighted case was considered in [112].

We say that the matrix $A = (a_{n,k})$ has almost monotone increasing (decreasing) rows if there is a constant K_1 (K_2), depending only on A, such that $a_{n,k} \leq K_1 a_{n,m}$ $(a_{n,m} \leq K_2 a_{n,k}$), where $0 \leq k \leq m \leq n$.

The main results presented in this section are following:

Theorem 4.1 ([123]). *Let* $f \in Lip(\alpha, p(\cdot), \omega)$, $0 < \alpha < 1$, $p(\cdot) \in \mathscr{P}_0(T)$, $\omega(\cdot) \in A_{p(\cdot)}(T)$ *and let* $A = (a_{n,k})$ *be a lower triangular matrix with* $\left|s_n^{(A)} - 1\right| = \mathscr{O}(n^{-\alpha})$. *If one of the conditions:*

$$\begin{aligned} (i)\ A \text{ has almost monotone decreasing rows and } (n+1)a_{n,0} &= \mathscr{O}(1), \\ (ii)\ A \text{ has almost monotone increasing rows and } (n+1)a_{n,r} &= \mathscr{O}(1) \end{aligned}$$

where r is the integer part of $/2$, *holds, then*

$$\left\| f - T_n^{(A)}(f)(x) \right\|_{p(\cdot),\omega} = \mathscr{O}(n^{-\alpha}).$$

Theorem 4.1 in the case of $\omega = 1$ was obtained in [111]. In the classical weighted Lebesgue spaces, when $1 < p < \infty$ and ω satisfies the classical Muckenhoupt condition A_p, it was proved in [112]. In the classical nonweighted Lebesgue spaces, under more restrictive conditions than (i) and (ii), i.e., in the case of monotone rows, the corresponding results were proved in [142].

Consider the lower triangular matrix $A = (a_{n,k})$ with $a_{n,k} = p_{n,k}/P_n$. Then $T_n^{(A)}(f)$ turns into the Nörlund mean $N_n(f)$, for which $s_n^{(A)} = 1$ and hence $\left|s_n^{(A)} - 1\right| = \mathscr{O}(n^{-\alpha})$. Let $0 < \alpha < 1$ and (p_n) be a sequence of positive numbers. If (p_n) is almost monotone decreasing, then the matrix A has almost monotone increasing rows and $(n+1)a_{n,r} \le K\dfrac{(n+1)p_r}{P_n} = K_1\dfrac{(r+1)p_r}{P_r} = \mathscr{O}(1)$, where $r = [n/2]$. Thus, A satisfies the condition (ii) of Theorem 4.1. If (p_n) is almost monotone increasing and $(n+1)p_n = \mathscr{O}(P_n)$, then A has almost monotone decreasing rows and $(n+1)a_{n,0} \le (n+1)\dfrac{p_n}{P_n} = \dfrac{1}{P_n}\mathscr{O}(P_n) = \mathscr{O}(1)$. Thus, A satisfies also the condition (i) of Theorem 4.1 and hence we have the following corollary:

Corollary 4.2. *Let* $f \in Lip(\alpha, p(\cdot), \omega)$, $0 < \alpha < 1$, $p(\cdot) \in \mathscr{P}_0(T)$, $\omega(\cdot) \in A_{p(\cdot)}(T)$ *and let* (p_n) *be a sequence of positive numbers. If one of the following conditions:*

(i) (p_n) *is almost monotone increasing and* $(n+1)p_n = \mathscr{O}(P_n)$,

(ii) (p_n) *is almost monotone decreasing,*

holds, then

$$\| f - N_n(f)(x) \|_{p(\cdot),\omega} = \mathscr{O}(n^{-\alpha}).$$

This result in the case of $\omega = 1$ and $p(\cdot) = p > 1$ was proved in [141], and earlier, under more restrictive conditions (in the case of monotone rows), in [101].

Theorem 4.3 ([123]). *Let* $f \in Lip(1, p(\cdot), \omega)$, $p(\cdot) \in \mathscr{P}_0(T)$, $\omega(\cdot) \in A_{p(\cdot)}(T)$ *and let* $A = (a_{n,k})$ *be a lower triangular matrix with* $\left|s_n^{(A)} - 1\right| = O(n^{-1})$. *If one of the following conditions:*

$$(i) \sum_{k=1}^{n-1} \left|a_{n,k-1} - a_{n,k}\right| = \mathscr{O}(n^{-1}),$$

$$(ii) \sum_{k=1}^{n-1} (n-k) \left|a_{n,k-1} - a_{n,k}\right| = \mathscr{O}(1),$$

holds, then

$$\left\|f - T_n^{(A)}(f)(x)\right\|_{p(\cdot),\omega} = \mathscr{O}(n^{-1}).$$

Theorem 4.3, in the nonweighted case and in the classical weighted Lebesgue spaces was proved in [111] and [112], respectively.

We consider Corollary 4.2 in the case of $\alpha = 1$. If

$$\sum_{k=1}^{n-1} |p_k - p_{k+1}| = \mathscr{O}(P_n/n), \tag{3.37}$$

then

$$\begin{aligned}\sum_{k=1}^{n-1} \left|a_{n,k-1} - a_{n,k}\right| &= \sum_{k=1}^{n-1} \left|\frac{p_{n-k+1}}{P_n} - \frac{p_{n-k}}{P_n}\right| \\ &= \frac{1}{P_n} \sum_{k=1}^{n-1} |p_k - p_{k+1}| = \frac{1}{P_n} \mathscr{O}(P_n/n)\end{aligned}$$

and hence the condition (3.37) implies the condition (i) of Theorem 4.3.

Now, from Theorem 4.3 we have

Corollary 4.4. *Let* $f \in Lip(1, p(\cdot), \omega)$, $p(\cdot) \in \mathscr{P}_0(T)$, $\omega(\cdot) \in A_{p(\cdot)}(T)$ *and let* (p_n) *be a sequence of positive numbers. If* $\sum_{k=1}^{n-1} |p_k - p_{k+1}| = \mathscr{O}(P_n/n)$ *,then*

$$\|f - N_n(f)\|_{p(\cdot),\omega} = \mathscr{O}(n^{-1}).$$

Chapter 4

Solvability in the Classical Sense of Fredholm Integro-differential Equations With Nonlinear Nonlocal Conditions

T.K. Yuldashev[1 *]
Zh.A. Artykova[2*]
[1]Tashkent State University of Economics, Tashkent, Uzbekistan
[2]Osh State University, Osh, Kyrgyzstan

Abstract

This paper addresses the classical solvability and construction of solutions for a nonlocal problem involving a second-order nonlinear Fredholm integro-differential equation with a degenerate kernel and two real parameters. The degenerate kernel method is utilized, and the nonlocal conditions are expressed as nonlinear functions. Specific challenges in constructing solutions, such as determining integration coefficients and source data, are analyzed. The parameters for which the classical solvability of the direct nonlocal problem is established are calculated, and corresponding solutions are derived. Methods of successive approximations and contraction mappings are employed to prove the uniqueness of solutions.

---

[*]Corresponding Author's Email: tursun.k.yuldashev@gmail.com, t.yuldashev@tsue.uz, jartykova@oshsu.kg

Keywords: Integro-differential equation, nonlinear integral conditions, degenerate kernel, classical solvability, regular and irregular parameter values.

AMS Subject Classification: 34B08, 34B10.

1. Problem statement

Integro-differential equations form a foundational component of mathematical physics and mechanics. These equations are widely used in studying problems across disciplines such as physics, mechanics, chemistry, economics, and other sciences. A significant body of literature has explored the properties of solutions to various types of integro-differential equations (see, e.g., [164–184]).

Integro-differential equations with a degenerate kernel have a prominent place in the theory of Fredholm integral and integro-differential equations. Ukrainian and Kazakh mathematicians have extensively studied first-order Fredholm integro-differential equations with degenerate kernels, where issues related to small denominators do not arise. Advancing the theory to include second-order Fredholm integro-differential equations with degenerate kernels and nonlocal conditions is a crucial development. Previous studies on Fredholm integro-differential equations with degenerate kernels can be found in [185–189].

This chapter investigates a nonlocal problem characterized by nonlinearity in the given integral conditions for a second-order nonlinear Fredholm integro-differential equation with a degenerate kernel and two real parameters. Regular and irregular parameter values are identified. We establish conditions for the nonexistence, uniqueness, and multiplicity of solutions for the nonlocal problem and construct the corresponding solutions.

On the interval $(0,T)$, we consider the following nonlinear Fredholm integro-differential equation:

$$u''(t)+\lambda^2 u(t)=\nu\int_0^T K(t,s)\left[su(s)+(T-s)u'(s)\right]ds+f(t,u(t)) \tag{4.1}$$

with the nonlinear boundary conditions

$$u(T)-\int_0^T su(s)ds=\varphi_1\left(\int_0^T H_1(s)u(s)ds\right), \tag{4.2}$$

$$u'(T)-\int_0^T (T-s)u'(s)ds=\varphi_2\left(\int_0^T H_2(s)u(s)ds\right), \tag{4.3}$$

where $T > \sqrt{2}$, λ is a positive parameter, ν is a nonzero real parameter, $0 \neq H_j(t) \in C[0,T]$, $j = 1,2$, $0 \neq f(t,u) \in C([0,T] \times R)$, $0 \neq K(t,s) = \sum_{i=1}^{k} a_i(t) b_i(s)$, $a_i(t), b_i(s) \in C[0,T]$. It is assumed that the functions $a_i(t)$ and $b_i(s)$ are linearly independent.

The positivity of the second term on the left-hand side of Equation (4.1) plays a critical role in determining solvability. If the second term were negative, the results obtained in this work would not hold.

Since the boundary conditions (4.2) and (4.3) are nonlinear, the nonlinear Fredholm integro-differential equation (4.1) does not have trivial solutions satisfying conditions (4.2) and (4.3). We investigate the existence and nonexistence of nontrivial solutions and determine the parameter values λ and ν for which the problem admits a unique solution or an infinite number of solutions. Additionally, we identify parameter values for which the problem has no solution. In this respect, the present work extends the findings of [190–193] and builds upon the works in [194, 195].

2. Integration of nonlinear boundary value problem

Given the degeneracy of the kernel in the Fredholm integro-differential equation (4.1), the equation can be rewritten as a nonhomogeneous second-order differential equation:

$$u''(t) + \lambda^2 u(t) = \nu \sum_{i=1}^{k} a_i(t) \tau_i + f(t, u(t)),$$

where

$$\tau_i = \int_0^T b_i(s) \left[s u(s) + (T - s) u'(s) \right] ds. \tag{4.4}$$

The general solution to this differential equation can be expressed in terms of sine and cosine functions:

$$u(t) = A_1 \cos \lambda t + A_2 \sin \lambda t + \frac{\nu}{\lambda} \sum_{i=1}^{k} \tau_i \int_0^t \sin \lambda (t - s) a_i(s) ds +$$

$$+ \frac{1}{\lambda} \int_0^t \sin \lambda (t - s) f(s, u(s)) ds, \tag{4.5}$$

where A_1, A_2 are yet unknown constants of integration. By differentiating (4.5), we find the derivative of this function:

$$u'(t) = -\lambda A_1 \sin\lambda t + \lambda A_2 \cos\lambda t + \nu \sum_{i=1}^{k} \tau_i \int_0^t \cos\lambda(t-s) a_i(s) ds +$$

$$+ \int_0^t \cos\lambda(t-s) f(s,u(s)) ds. \tag{4.6}$$

To determine the unknown coefficients A_1 and A_2 in Equation (4.6), we substitute the representations from Equations (4.5) and (4.6) into the nonlinear integral conditions (4.2) and (4.3). This leads to the following transcendental system of linear algebraic equations:

$$\begin{cases} A_1\chi_{11}(\lambda) + A_2\chi_{12}(\lambda) = \chi_{13}(\lambda,u), \\ A_1\chi_{21}(\lambda) + A_2\chi_{22}(\lambda) = \chi_{23}(\lambda,u), \end{cases} \tag{4.7}$$

where

$$\chi_{11}(\lambda) = \cos\lambda T - \frac{T}{\lambda}\sin\lambda T + \frac{1}{\lambda^2}(1-\cos\lambda T),\ \chi_{12}(\lambda) = \sin\lambda T + \frac{T}{\lambda}\cos\lambda T - \frac{1}{\lambda^2}\sin\lambda T,$$

$$\chi_{21}(\lambda) = -\lambda\sin\lambda T - T - \frac{1}{\lambda}\sin\lambda T,\ \chi_{22}(\lambda) = \lambda\cos\lambda T - \frac{1}{\lambda}(1-\cos\lambda T),$$

$$\chi_{13}(\lambda,u) = \varphi_1\left(\int_0^T H_1(s)u(s)ds\right) - \eta(T,\lambda,f) + \int_0^T s\cdot\eta(s,\lambda,f)ds,$$

$$\chi_{23}(\lambda,u) = \varphi_2\left(\int_0^T H_2(s)u(s)ds\right) - \eta'(T,\lambda,f) + \int_0^T (T-s)\cdot\eta'(s,\lambda,f)ds,$$

$$\eta(t,\lambda,f) = \frac{\nu}{\lambda}\sum_{i=1}^{k}\tau_i h_i(t,\lambda) + \frac{1}{\lambda}\int_0^t \sin\lambda(t-s) f(s,u(s))ds,$$

$$h_i(t,\lambda) = \int_0^t \sin\lambda(t-s)a_i(s)ds, \quad i=\overline{1,k}.$$

For the system of linear equations (4.7), we calculate parameter values λ under two distinct cases:

1) $\chi_{11}(\lambda) = \chi_{12}(\lambda) = 0$,
2) $\chi_{21}(\lambda) = \chi_{22}(\lambda) = 0$.

If the sets of such λ values are nonempty, they are excluded from further consideration. Because we do not put $\chi_{j3}(\lambda, u) = 0, \ j = 1, 2$.

Then we calculate the values of parameter λ in the case
3) $\chi_{11}(\lambda)\chi_{22}(\lambda) - \chi_{12}(\lambda)\chi_{21}(\lambda) = 0$.
In this case the uniqueness of the solutions of the problem is violated.

We find out that in case

$$\chi_{11}(\lambda)\chi_{22}(\lambda) - \chi_{12}(\lambda)\chi_{21}(\lambda) \neq 0$$

the uniqueness of the solution of the system (4.7) is not violated. So, in this case, we continue the process of studying the existence and uniqueness of the solution of the problem (4.1)–(4.3) and constructing this solution.

We graphically study the solvability of the following four transcendental equations: $\chi_{ij}(\lambda) = 0, \ i, j = 1, 2$.

1. The set of solutions of the equation $\chi_{11}(\lambda) = 0$ coincides with the solutions of the equation

$$(\lambda^2 - 2)\tan^2 \frac{T}{2}\lambda + 2T\lambda \cdot \tan \frac{T}{2}\lambda - \lambda^2 = 0, \ \lambda > 0. \tag{4.8}$$

The solvability of the transcendental equation (4.8) is illustrated in Figure 1 for $T = 2$.

2. The set of solutions of the second transcendental equation $\chi_{12}(\lambda) = 0$ is equal to the set of solutions of the following equation:

$$\tan T\lambda - \frac{T\lambda}{1 - \lambda^2} = 0, \ \lambda > 0. \tag{4.9}$$

The solvability of the equation (4.9) is illustrated in Figure 2 for $T = 2$.

3. Now we consider the third transcendental equation $\chi_{21}(\lambda) = 0$. The set of solutions of this equation is equivalent to the set of solutions of the following equation:

$$\sin T\lambda + \frac{T\lambda}{1 + \lambda^2} = 0, \ \lambda > 0. \tag{4.10}$$

The solvability of the equation (4.10) is illustrated in the Figure 3 for $T = 2$.

4. The set of solutions of the fourth equation $\chi_{22}(\lambda) = 0$ is equivalent to the set of solutions of the equation:

$$\cos T\lambda - \frac{1}{1 + \lambda^2} = 0, \ \lambda > 0. \tag{4.11}$$

The solvability of the equation (4.11) is illustrated in the Figure 4 for $T = 2$.

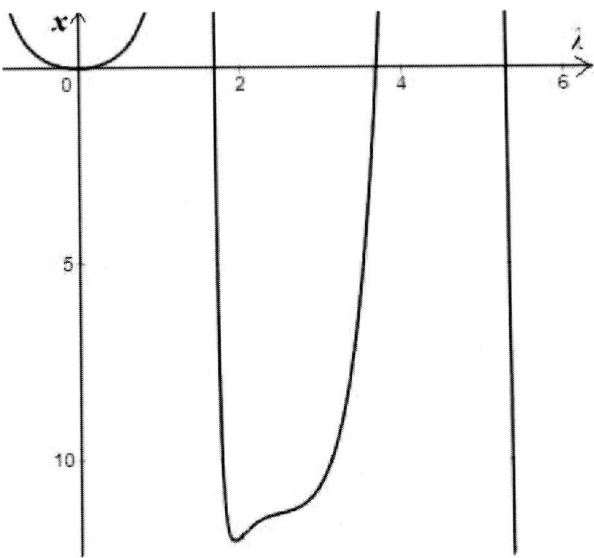

Figure 4.1. Graph of the function in (4.8). T=2

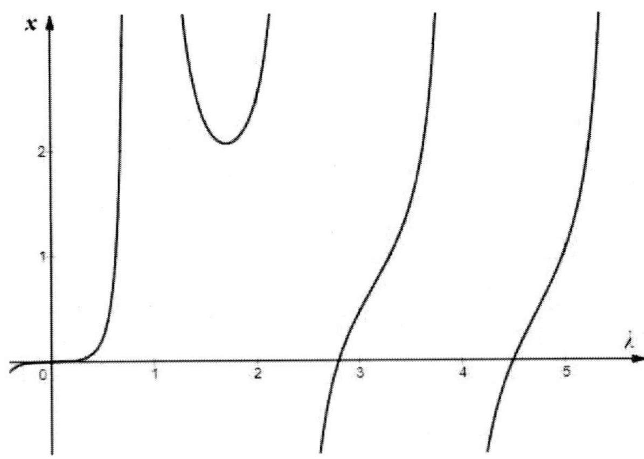

Figure 4.2. Graph of the function in (4.9). T=2

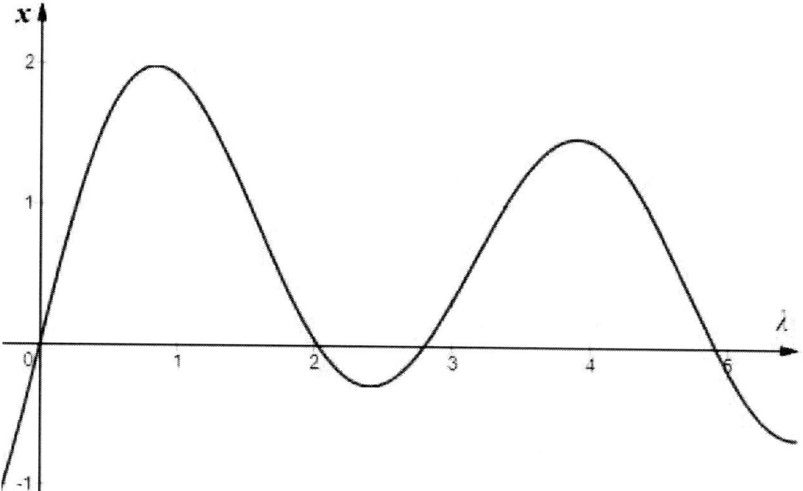

Figure 4.3. Graph of the function in (4.10). T=2

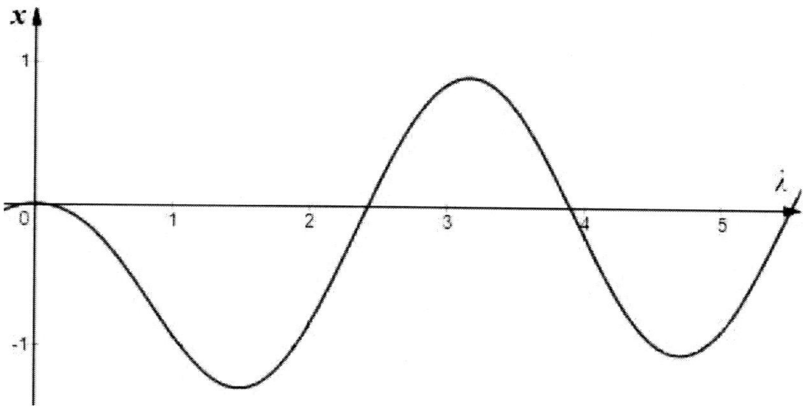

Figure 4.4. Graph of the function in (4.11). T=2

We denote the sets of solutions of transcendental equations (4.8)–(4.11) by $\Im_i (i=1,2,3,4)$, respectively. So, we have a notation $\Lambda_i = (0,\infty) \setminus \Im_i$, $i = 1,2,3,4$.

Thus, we have established that every transcendental equation $\chi_{ij}(\lambda) = 0$, $i,j = 1,2$ has solutions with respect to a parameter λ. Interesting cases are when the parameter values simultaneously satisfy the two transcendental equations discussed above.

Case 2.1. It is clear from the system (4.7) that consideration of the case $\chi_{11}(\lambda) = \chi_{12}(\lambda) = 0$ is natural. Let us determine the values of the parameter λ, for that the case 2.1 takes place. Indeed, if this case takes place, then the algebraic equation:

$$\chi_{11}^2(\lambda) + \chi_{12}^2(\lambda) = 0 \tag{4.12}$$

has a solution. Equation (4.12) is equivalent to the following transcendental equation:

$$\cos(\lambda T + \phi) + \frac{\sqrt{(\lambda^2-1)^2 + \lambda^2 T^2}}{2} + \frac{1}{2\sqrt{(\lambda^2-1)^2 + \lambda^2 T^2}} = 0, \tag{4.13}$$

where $\phi = \arccos \dfrac{\lambda^2 - 1}{\sqrt{(\lambda^2-1)^2 + \lambda^2 T^2}}$. Equation (4.13) has a solution if the following inequality is valid:

$$\frac{\sqrt{(\lambda^2-1)^2 + \lambda^2 T^2}}{2} + \frac{1}{2\sqrt{(\lambda^2-1)^2 + \lambda^2 T^2}} < 1.$$

Indeed, if we assume the opposite:

$$\frac{\sqrt{(\lambda^2-1)^2 + \lambda^2 T^2}}{2} + \frac{1}{2\sqrt{(\lambda^2-1)^2 + \lambda^2 T^2}} > 1,$$

then we obtain $\lambda > \sqrt{2 - T^2}$. We have arrived at an incorrect inequality. Since the parameter λ is positive and $T > \sqrt{2}$, then the inequality $\lambda > \sqrt{2 - T^2}$ does not make sense in the set of real numbers. So, we have a contradiction. Consequently, equation (4.13) has a solution. In order to verify this fact, we construct a graph of the function (4.13) in coordinate plane for $T = 2$. From Figure 5, it can be seen that this case is possible only for small values of the parameter λ.

Denote the set of values of the parameter λ, for which the equation (4.12) has a solution, by $\Im_5$. The linear system (4.7) can have any solution, if $\chi_{13}(\lambda, u) = 0$ for values of the parameter $\lambda \in \Im_5$. It is possible if and only if the following

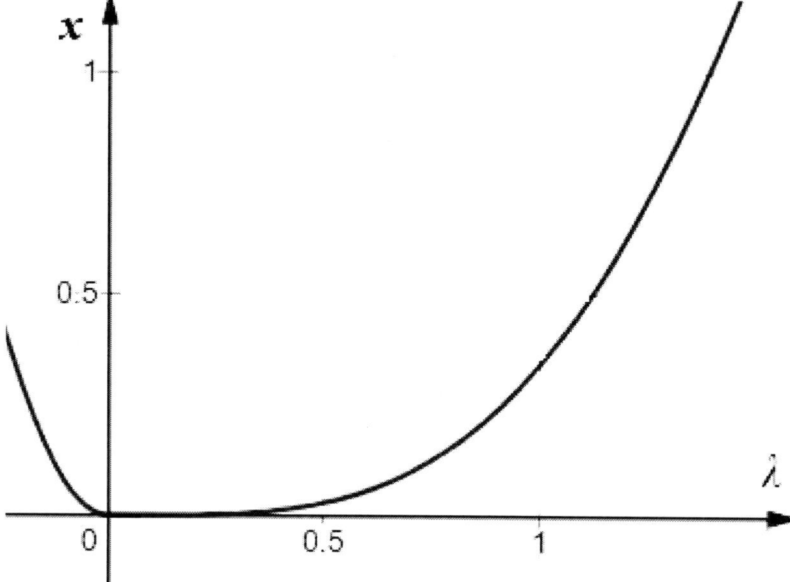

Figure 4.5. Graph of the function in (4.13). T=2

equality is valid:

$$\varphi_1\left(\int_0^T H_1(s)u(s)ds\right) = \eta(T,\lambda,f) + \int_0^T s\cdot\eta(s,\lambda,f)ds.$$

However, this equality does not work in our favor. So, we set

$$\varphi_1\left(\int_0^T H_1(s)u(s)ds\right) \neq \eta(T,\lambda,f) + \int_0^T s\cdot\eta(s,\lambda,f)ds. \tag{4.14}$$

Consequently, in this case, the system (4.7) has no solution for $\lambda \in \Im_5$. Note that $\Im_5 = \Im_1 \cap \Im_2$. Therefore, $\Im_5 \not\subset \Lambda_1$ and $\Im_5 \not\subset \Lambda_2$.

Case 2.2: $\chi_{21}(\lambda) = \chi_{22}(\lambda) = 0$. In this case we will calculate the set of values of the parameter λ. Determine the values of the parameter λ, for which the following algebraic equation has a solution:

$$\chi_{21}^2(\lambda) + \chi_{22}^2(\lambda) = 0. \tag{4.15}$$

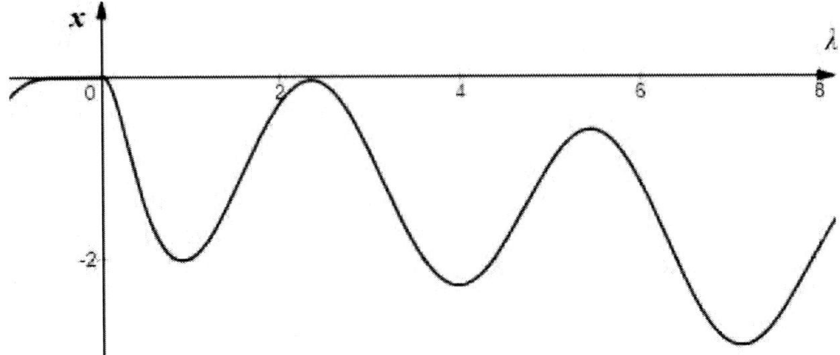

Figure 4.6. Graph of the function in (4.16). T=2

Equation (4.15) is equivalent to the following transcendental equation:

$$\cos(\lambda T+\phi)-\frac{(\lambda^2+1)^2+1+\lambda^2T^2}{2(\lambda^2+1)\sqrt{1+\lambda^2T^2}}=0,\ \ \lambda>0, \tag{4.16}$$

where $\phi=\arccos\dfrac{1}{1+\lambda^2T^2}$.

The equation (4.16) has a solution, if the inequality

$$\frac{(\lambda^2+1)^2+1+\lambda^2T^2}{2(\lambda^2+1)\sqrt{1+\lambda^2T^2}}<1,$$

holds. If we assume the opposite of the inequality above:

$$\frac{(\lambda^2+1)^2+1+\lambda^2T^2}{2(\lambda^2+1)\sqrt{1+\lambda^2T^2}}>1,$$

then we obtain $\lambda^2>T^2-2$. Since $T>\sqrt{2}$, we obtain correct inequality. Hence, it is clear that the equation (4.16) has no solution. Therefore, the set of values of λ, for which $\chi_{21}(\lambda)=\chi_{22}(\lambda)=0$, is empty. This statement can also be seen from Figure 6. So, this case 2.2 is impossible.

Case 2.3. This case is where the main determinant of the system vanishes: $\chi_{11}(\lambda)\cdot\chi_{22}(\lambda)-\chi_{12}(\lambda)\cdot\chi_{21}(\lambda)=0$. Let us determine the values of λ in this case. Indeed, if this case takes place, then the algebraic equation

$$\chi_{11}(\lambda)\cdot\chi_{22}(\lambda)-\chi_{12}(\lambda)\cdot\chi_{21}(\lambda)=0 \tag{4.17}$$

has a solution. Equation (4.17) is equivalently reduced to the following transcendental equation:

$$(2+T^2\lambda^2)\cos T\lambda+\lambda^3T\sin T\lambda+\lambda^4-2=0,\ \ \lambda>0. \tag{4.18}$$

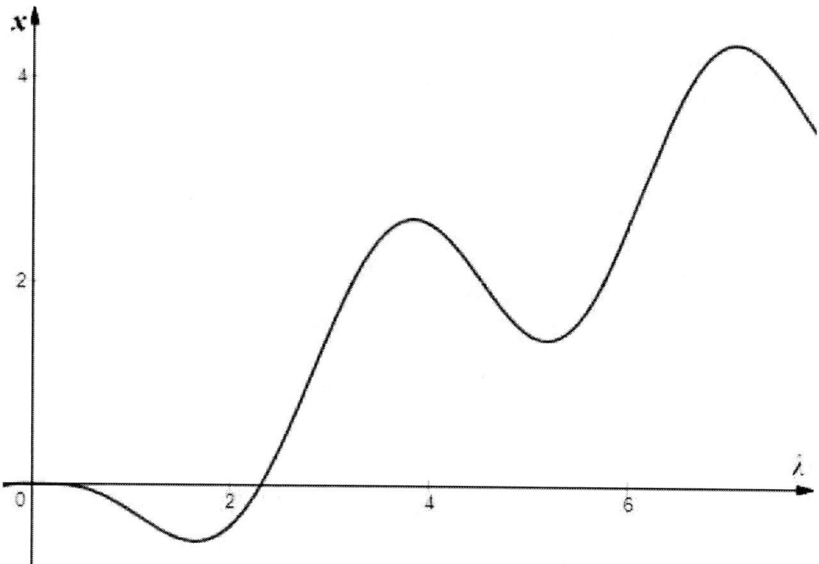

Figure 4.7. Graph of the function in (4.19). T=2

One can rewrite the transcendental equation (4.18) as

$$\cos(\lambda T - \theta) - \frac{2-\lambda^4}{\sqrt{(2+\lambda^2 T^2)^2 + \lambda^6 T^2}} = 0, \tag{4.19}$$

where

$$\theta = \arccos \frac{2+T^2\lambda^2}{\sqrt{(2+T^2\lambda^2)^2 + \lambda^6 T^2}}.$$

The graph of the function in (4.19) is illustrated by Figure 7 for $T = 2$. From Figure 7 it is clear that the case 2.3 is possible for small values of parameter λ and for a value on the interval $(2, 2.5)$.

The set of values of the parameter λ, for which equation (4.19) has solutions will be denoted by $\Im_6$. Using this notation, we adopt a new one: $\Lambda_5 = \cup_{i=1}^{4} \Lambda_i \setminus \Im_6$. For values of the parameter λ from the set Λ_5, we study the influence of the second parameter ν on the solvability of nonlocal nonlinear problem (4.1)–(4.3). For certain values of the parameter ν, we construct a solution. The values of the parameter $\lambda \in \Lambda_5$ are called regular.

3. Regular values of the parameter λ

We consider the regular values of the parameter $\lambda \in \Lambda_5$. In this case we have $D(\lambda) = \chi_{11}(\lambda)\chi_{22}(\lambda) - \chi_{12}(\lambda)\chi_{21}(\lambda) \neq 0$. Then solving the linear system of equations (4.7) by the Cramer method, uniquely determine A_1 and A_2:

$$A_1 = \frac{1}{D(\lambda)} \begin{vmatrix} \chi_{13}(\lambda,u) & \chi_{12}(\lambda) \\ \chi_{23}(\lambda,u) & \chi_{22}(\lambda) \end{vmatrix}, \tag{4.20}$$

$$A_2 = \frac{1}{D(\lambda)} \begin{vmatrix} \chi_{11}(\lambda) & \chi_{13}(\lambda,u) \\ \chi_{21}(\lambda) & \chi_{23}(\lambda,u) \end{vmatrix}. \tag{4.21}$$

Substituting (4.20) and (4.21) into representations (4.5) and (4.6) and using the notations

$$\chi_{13}(\lambda,u) = \varphi_1\left(\int_0^T H_1(s)u(s)ds\right) - \eta(T,\lambda,f) + \int_0^T s\cdot\eta(s,\lambda,f)ds,$$

$$\chi_{23}(\lambda,u) = \varphi_2\left(\int_0^T H_2(s)u(s)ds\right) - \eta'(T,\lambda,f) + \int_0^T (T-s)\cdot\eta'(s,\lambda,f)ds,$$

$$\eta(t,\lambda,f) = \frac{\nu}{\lambda}\sum_{i=1}^k \tau_i h_i(t,\lambda) + \frac{1}{\lambda}\int_0^t \sin\lambda(t-s) f(s,u(s))ds,$$

$$h_i(t,\lambda) = \int_0^t \sin\lambda(t-s)a_i(s)ds,\ i=\overline{1,k},$$

we obtain the representations for formal solution of the problem and for its derivative:

$$u(t,\lambda) = \varphi_1\left(\int_0^T H_1(s)u(s)ds\right)\delta_{11}(t,\lambda) -$$

$$-\varphi_2\left(\int_0^T H_2(s)u(s)ds\right)\delta_{12}(t,\lambda) + \frac{\nu}{\lambda}\sum_{i=1}^k \tau_i\delta_{13i}(t,\lambda) +$$

$$+\int_0^T f(s,u(s))\delta_{14}(t,s,\lambda)ds + \int_0^T\int_0^s f(\theta,u(\theta))\delta_{15}(t,s,\theta,\lambda)d\theta ds,\ \lambda\in\Lambda_5, \tag{4.22}$$

$$u'(t,\lambda) = -\varphi_1 \left(\int_0^T H_1(s)u(s)ds \right) \delta_{21}(t,\lambda) +$$

$$+\varphi_2 \left(\int_0^T H_2(s)u(s)ds \right) \delta_{22}(t,\lambda) - \frac{\nu}{\lambda} \sum_{i=1}^{k} \tau_i \delta_{23i}(t,\lambda) +$$

$$+\int_0^T f(s,u(s))\delta_{24}(t,s,\lambda)ds - \int_0^T \int_0^s f(\theta,u(\theta))\delta_{25}(t,s,\theta,\lambda)d\theta ds, \ \lambda \in \Lambda_5, \tag{4.23}$$

where

$$\delta_{11}(t,\lambda) = \delta_{11}(\lambda)\cos\lambda t - \delta_{21}(\lambda)\sin\lambda t, \ \delta_{12}(t,\lambda) = \delta_{12}(\lambda)\cos\lambda t - \delta_{22}(\lambda)\sin\lambda t,$$

$$\delta_{13i}(t,\lambda) = \delta_{13i}(\lambda)\cos\lambda t - \delta_{23i}(\lambda)\sin\lambda t + h_i(t,\lambda), \ i = \overline{1,k},$$

$$\delta_{14}(t,s,\lambda) = \begin{cases} \delta_{14}(s,\lambda)\cos\lambda t + \delta_{24}(s,\lambda)\sin\lambda t, \ t \le s \le T, \\ \dfrac{1}{\lambda}\sin\lambda(t-s) + \delta_{14}(s,\lambda)\cos\lambda t + \delta_{24}(s,\lambda)\sin\lambda t, \ 0 \le s \le t, \end{cases}$$

$$\delta_{15}(t,s,\theta,\lambda) = \delta_{15}(s,\theta,\lambda)\cos\lambda t - \delta_{25}(s,\theta,\lambda)\sin\lambda t,$$

$$\delta_{21}(t,\lambda) = \lambda\delta_{11}(\lambda)\sin\lambda t + \lambda\delta_{21}(\lambda)\cos\lambda t, \delta_{22}(t,\lambda) = \lambda\delta_{12}(\lambda)\sin\lambda t + \lambda\delta_{22}(\lambda)\cos\lambda t,$$

$$\delta_{23i}(t,\lambda) = \lambda\delta_{13i}(\lambda)\sin\lambda t + \lambda\delta_{23i}(\lambda)\cos\lambda t - h_i'(t,\lambda), \ i = \overline{1,k},$$

$$\delta_{24}(t,s,\lambda) = \begin{cases} -\lambda\delta_{14}(s,\lambda)\sin\lambda t + \lambda\delta_{24}(s,\lambda)\cos\lambda t, \ t \le s \le T, \\ -\cos\lambda(t-s) - \lambda\delta_{14}(s,\lambda)\sin\lambda t + \lambda\delta_{24}(s,\lambda)\cos\lambda t, \ 0 \le s \le t, \end{cases}$$

$$\delta_{25}(t,s,\theta,\lambda) = \lambda\delta_{15}(s,\theta)\sin\lambda t + \lambda\delta_{25}(s,\theta)\cos\lambda t,$$

$$\delta_{11}(\lambda) = \frac{\chi_{22}(\lambda)}{D(\lambda)}, \ \delta_{12}(\lambda) = \frac{\chi_{12}(\lambda)}{D(\lambda)},$$

$$\delta_{j3i}(\lambda) = \left[\int_0^T s \cdot h_i(s,\lambda)ds - h_i(T,\lambda) \right] \sigma_{j1}(\lambda) - \left[\int_0^T (T-s)h_i'(s,\lambda)ds - h_i'(T,\lambda) \right] \sigma_{j2}(\lambda),$$

$$j = 1,2, \ i = \overline{1,k}, \ \sigma_{21}(\lambda) = \frac{\chi_{21}(\lambda)}{D(\lambda)}, \ \sigma_{22}(\lambda) = \frac{\chi_{11}(\lambda)}{D(\lambda)},$$

$$\sigma_{j4}(t,\lambda) = \frac{1}{\lambda}\sigma_{j1}(\lambda)\sin\lambda(T-t) + \sigma_{j2}(\lambda)\cos\lambda(T-t), \ j = 1,2,$$

$$\sigma_{j5}(t,s,\lambda) = \frac{1}{\lambda}\sigma_{j1}(\lambda)t\sin\lambda(t-s) + \sigma_{j2}(\lambda)(T-t)\cos\lambda(t-s), \ j = 1,2.$$

Functions (4.22) and (4.23) contain still unknown quantities τ_i. We need to determine these quantities. So, substituting the representations (4.22) and (4.23) into (4.4), we arrive at a linear system of algebraic equations (SAE)

$$\tau_i - \frac{\nu}{\lambda} \sum_{j=1}^{k} \tau_j \Phi_{ij}(\lambda) = \varphi_1 \Psi_{1i}(\lambda) + \varphi_2 \Psi_{2i}(\lambda) + \Psi_{3i}(f,\lambda), \ i = \overline{1,k}, \ \lambda \in \Lambda_5, \tag{4.24}$$

where

$$\Phi_{ij}(\lambda)=\int_0^T b_i(s)\left[s\cdot\delta_{13j}(s,\lambda)-(T-s)\delta_{23j}(s,\lambda)\right]ds,\ \ i,j=\overline{1,k},$$

$$\Psi_{1i}(\lambda)=\int_0^T b_i(s)\left[s\cdot\delta_{11}(s,\lambda)-(T-s)\delta_{21}(s,\lambda)\right]ds,\ \ i=\overline{1,k},\tag{4.25}$$

$$\Psi_{2i}(\lambda)=\int_0^T b_i(s)\left[-s\cdot\delta_{12}(s,\lambda)+(T-s)\delta_{22}(s,\lambda)\right]ds,\ \ i=\overline{1,k},\tag{4.26}$$

$$\Psi_{3i}(f,\lambda)=\int_0^T b_i(s)\int_0^T f(\theta,u(\theta))\left[\theta\cdot\delta_{14}(\theta,\lambda)+(T-\theta)\delta_{24}(\theta,\lambda)\right]d\theta ds+$$

$$+\int_0^T b_i(s)\int_0^T\int_0^\theta f(\xi,u(\xi))\left[\xi\cdot\delta_{15}(\theta,\xi,\lambda)+(T-\xi)\delta_{25}(\theta,\xi,\lambda)\right]d\xi d\theta ds,\tag{4.27}$$

$i=\overline{1,k}$.

As we now from linear algebra, the SAE (4.24) is uniquely solvable for any finite right-hand side Ψ_{mi}, if the following Fredholm condition is fulfilled:

$$\Delta_\Phi(\nu,\lambda)=\begin{vmatrix} 1-\frac{\nu}{\lambda}\Phi_{11}(\lambda) & \frac{\nu}{\lambda}\Phi_{12}(\lambda) & \dots & \frac{\nu}{\lambda}\Phi_{1k}(\lambda)\\ 1-\frac{\nu}{\lambda}\Phi_{21}(\lambda) & \frac{\nu}{\lambda}\Phi_{22}(\lambda) & \dots & \frac{\nu}{\lambda}\Phi_{2k}(\lambda)\\ \dots & \dots & \dots & \dots\\ \frac{\nu}{\lambda}\Phi_{k1}(\lambda) & \Phi_{k2}(\lambda) & \dots & 1-\frac{\nu}{\lambda}\Phi_{kk}(\lambda)\end{vmatrix}\neq 0.\tag{4.28}$$

The determinant $\Delta_\Phi(\nu,\lambda)$ in (4.28) is a polynomial with respect to $\frac{\nu}{\lambda}$ of degree at most k. The equation $\Delta_\Phi(\nu,\lambda)=0$ has at most k distinct real roots. We denote them by μ_r, $1\leq r\leq k$. Then $\nu=\lambda\,\mu_r$ are the eigenvalues of kernel of the Fredholm integro-differential equation (4.1). For other values of the parameter $\nu\neq\lambda\,\mu_r$, the condition $\Delta_\Phi(\nu,\,\lambda)\neq 0$ holds.

Consider the following two sets of values of parameters:

$$\Omega_5=\left\{\lambda\in\Lambda_5,\ \nu=\lambda\mu_r\right\},\ \tilde{\Omega}_5=\left\{\lambda\in\Lambda_5,\ \nu\neq\lambda\mu_r\right\}.$$

In case of $\tilde{\Omega}_5(\nu,\lambda)$, the solution of SAE (4.24) has the form

$$\tau_i=\varphi_1\frac{\Delta_{\Psi_{1i}}(\nu,\lambda)}{\Delta_\Phi(\nu,\lambda)}+\varphi_2\frac{\Delta_{\Psi_{2i}}(\nu,\lambda)}{\Delta_\Phi(\nu,\lambda)}+\frac{\Delta_{\Psi_{3i}}(f,\nu,\lambda)}{\Delta_\Phi(\nu,\lambda)},\ \ i=\overline{1,k},\tag{4.29}$$

where $\Delta_{\Psi_{mi}}(\nu,\lambda) =$

$$= \begin{vmatrix} 1-\frac{\nu}{\lambda}\Phi_{11}(\lambda) & \ldots & \frac{\nu}{\lambda}\Phi_{1(j-1)}(\lambda) & \Psi_{m1i}(\lambda) & \frac{\nu}{\lambda}\Phi_{1(j+1)}(\lambda) & \ldots & \frac{\nu}{\lambda}\Phi_{1k}(\lambda) \\ \frac{\nu}{\lambda}\Phi_{21}(\lambda) & \ldots & \frac{\nu}{\lambda}\Phi_{2(j-1)}(\lambda) & \Psi_{m2i}(\lambda) & \frac{\nu}{\lambda}\Phi_{2(j+1)}(\lambda) & \ldots & \frac{\nu}{\lambda}\Phi_{2k}(\lambda) \\ \ldots & \ldots & \ldots & \ldots & \ldots & \ldots & \ldots \\ \frac{\nu}{\lambda}\Phi_{k1}(\lambda) & \ldots & \frac{\nu}{\lambda}\Phi_{k(j-1)}(\lambda) & \Psi_{mki}(\lambda) & \frac{\nu}{\lambda}\Phi_{k(j+1)}(\lambda) & \ldots & 1-\frac{\nu}{\lambda}\Phi_{kk}(\lambda) \end{vmatrix},$$

$m = 1, 2, 3.$

Substituting (4.29) into (4.22), we derive a new representation for solution of the problem (4.1)–(4.3) in the case $(\lambda,\nu)\in\tilde{\Omega}_5$:

$$u(t,\lambda,\nu) = I(t;u) \equiv \varphi_1\left(\int_0^T H_1(s)u(s)ds\right)V_1(t,\lambda,\nu)+$$

$$+\varphi_2\left(\int_0^T H_2(s)u(s)ds\right)V_2(t,\lambda,\nu)+\frac{\nu}{\lambda}\sum_{i=1}^{k}\frac{\Delta_{\Psi_{3i}}(f,\nu,\lambda)}{\Delta_{\Phi_i}(\nu,\lambda)}\delta_{13i}(t,\lambda)+$$

$$+\int_0^T f(s,u(s))\delta_{14}(t,s,\lambda)ds+\int_0^T\int_0^s f(\theta,u(\theta))\delta_{15}(t,s,\theta,\lambda)d\theta ds, \quad (4.30)$$

where

$$V_1(t,\lambda,\nu) = \delta_{11}(t,\lambda)+\frac{\nu}{\lambda}\sum_{i=1}^{k}\frac{\Delta_{\Psi_{1i}}(\nu,\lambda)}{\Delta_{\Phi}(\nu,\lambda)}\delta_{13i}(t,\lambda),$$

$$V_2(t,\lambda,\nu) = -\delta_{12}(t,\lambda)+\frac{\nu}{\lambda}\sum_{i=1}^{k}\frac{\Delta_{\Psi_{2i}}(\nu,\lambda)}{\Delta_{\Phi}(\nu,\lambda)}\delta_{13i}(t,\lambda).$$

4. Unique solvability of the equation (4.30)

In this section we prove that the equation (4.30) has a unique solution for the values of parameters $(\lambda,\nu)\in\tilde{\Omega}_5$. To do so, we need to consider a Banach space of continuous functions $C[0,T]$ and equip with the following norm:

$$\|u(t)\| = \max_{0\le t\le T}|u(t)|.$$

Let the following conditions be fulfilled:

$$|f(t,0)|\le M_1(t),\quad 0<M_1(t)\in C[0,T]; \quad (4.31)$$

$$|f(t,u_1(t))-f(t,u_2(t))|\le L(t)\,|u_1(t)-u_2(t)|,\quad 0<L(t)\in C[0,T]. \quad (4.32)$$

$$\left|\varphi_j(0)\right| \le M_{j+1},\ \ 0<M_{j+1}=\text{const},\ j=1,2; \tag{4.33}$$

$$\left|\varphi_j\left(\int\limits_0^T H_j(s)\,u_1(s)\,ds\right)-\varphi_j\left(\int\limits_0^T H_j(s)\,u_2(s)\,ds\right)\right| \le$$

$$\le L_0\int\limits_0^T \left|H_j(s)\right|\cdot\left|u_1(s)-u_2(s)\right|ds, \tag{4.34}$$

where

$$0<L_0=\text{const},\ \ \int\limits_0^T \left|H_j(s)\right|ds<\infty,\ \ j=1,2.$$

We define the iteration process of approximations for the values of parameters $(\lambda,\nu)\in\tilde{\Omega}_5$ as

$$\begin{cases} u_0(t,\lambda,\nu)=0, \\ u_{n+1}(t,\lambda,\nu)=I(t;u_n),\ \ n=0,1,2,\ldots \end{cases} \tag{4.35}$$

Taking the function (4.27) and the condition (4.31) into account, for the quantity $\Delta_{\Psi_{3i}}(f_0,\nu,\lambda)$, we obtain the estimate

$$\left|\Delta_{\Psi_{3i}}(f_0,\nu,\lambda)\right| \le (N_1+N_2)\left|\bar{\Delta}_{\Psi_{3i}}(\nu,\lambda)\right|,\ \ i=\overline{1,k},$$

where

$$N_1=\int\limits_0^T M(t)\left|t\cdot\delta_{14}(t,\lambda)+(T-t)\,\delta_{24}(t,\lambda)\right|dt,$$

$$N_2=\int\limits_0^T\int\limits_0^t M(s)\left|s\cdot\delta_{15}(t,s,\lambda)+(T-s)\,\delta_{25}(t,s,\lambda)\right|ds\,dt,$$

$$\bar{\Delta}_{\Psi_{3i}}(\nu,\lambda)=\begin{vmatrix} 1-\frac{\nu}{\lambda}\Phi_{11}(\lambda) & \ldots & \frac{\nu}{\lambda}\Phi_{1(j-1)}(\lambda) & \bar{\Psi}_{31i} & \frac{\nu}{\lambda}\Phi_{1(j+1)}(\lambda) & \ldots & \frac{\nu}{\lambda}\Phi_{1k}(\lambda) \\ \frac{\nu}{\lambda}\Phi_{21}(\lambda) & \ldots & \frac{\nu}{\lambda}\Phi_{2(j-1)}(\lambda) & \bar{\Psi}_{32i} & \frac{\nu}{\lambda}\Phi_{2(j+1)}(\lambda) & \ldots & \frac{\nu}{\lambda}\Phi_{2k}(\lambda) \\ \ldots & \ldots & \ldots & \ldots & \ldots & \ldots & \ldots \\ \frac{\nu}{\lambda}\Phi_{k1}(\lambda) & \ldots & \frac{\nu}{\lambda}\Phi_{k(j-1)}(\lambda) & \bar{\Psi}_{3ki} & \frac{\nu}{\lambda}\Phi_{k(j+1)}(\lambda) & \ldots & 1-\frac{\nu}{\lambda}\Phi_{kk}(\lambda) \end{vmatrix},$$

$$\bar{\Psi}_{3i}=\int\limits_0^T \left|b_i(s)\right|ds,\ \ i=\overline{1,k}.$$

Since the functions $V_1(t,\lambda,\nu)$ and $V_2(t,\lambda,\nu)$ are continuous on the segment $[0,T]$, by virtue of last estimate and conditions (4.31) and (4.33), from approximations (4.35) we obtain

$$\|u_1(t,\lambda,\nu)-u_0(t,\lambda,\nu)\| \le |\varphi_1(0)|\max_{0\le t\le T}|V_1(t,\lambda,\nu)|+|\varphi_2(0)|\max_{0\le t\le T}|V_2(t,\lambda,\nu)|+$$

$$+\frac{|\nu|}{\lambda}\sum_{i=1}^{k}\frac{\left|\Delta_{\Psi_{3i}}(f_0,\nu,\lambda)\right|}{|\Delta_{\Phi}(\nu,\lambda)|}\max_{0\leq t\leq T}|\delta_{13i}(t,\lambda)|+$$

$$+\max_{0\leq t\leq T}\int_0^T|f(s,u_0(s))|\cdot|\delta_{14}(t,s,\lambda)|ds+\max_{0\leq t\leq T}\int_0^T\int_0^s|f(\theta,u_0(\theta))|\cdot|\delta_{15}(t,s,\theta,\lambda)|d\theta ds\leq$$

$$\leq(M_2+M_3)V_0+(N_1+N_2)\frac{|\nu|}{\lambda}\sum_{i=1}^{k}\frac{\left|\bar{\Delta}_{\Psi_{3i}}(\nu,\lambda)\right|}{|\Delta_{\Phi}(\nu,\lambda)|}\max_{0\leq t\leq T}|\delta_{13i}(t,\lambda)|+$$

$$+\max_{0\leq t\leq T}\int_0^T|M_1(s)|\cdot|\delta_{14}(t,s,\lambda|ds+$$

$$+\max_{0\leq t\leq T}\int_0^T\int_0^s|M_1(\theta)|\cdot|\delta_{15}(t,s,\theta,\lambda|d\theta ds<\infty,\tag{4.36}$$

where $V_0=\max\left\{\max_{0\leq t\leq T}|V_1(t,\lambda,\nu)|;\ \max_{0\leq t\leq T}|V_2(t,\lambda,\nu)|\right\}$.

As in the case of estimate above, we have

$$\left|\Delta_{\Psi_{3i}}(f_n,\nu,\lambda)-\Delta_{\Psi_{3i}}(f_{n-1},\nu,\lambda)\right|\leq$$

$$\leq(\bar{N}_1+\bar{N}_2)\left|\bar{\Delta}_{\Psi_{3i}}(\nu,\lambda)\right|\cdot\|u_n(t,\lambda,\nu)-u_{n-1}(t,\lambda,\nu)\|,\ \ i=\overline{1,k},\tag{4.37}$$

where

$$\bar{N}_1=\int_0^T L(t)\,|t\cdot\delta_{14}(t,\lambda)+(T-t)\,\delta_{24}(t,\lambda)|\,dt,$$

$$\bar{N}_2=\int_0^T\int_0^t L(s)\,|s\cdot\delta_{15}(t,s,\lambda)+(T-s)\,\delta_{25}(t,s,\lambda)|\,ds\,dt.$$

Similarly, by virtue of (4.32), (4.34) and (4.37), we obtain

$$\|u_{n+1}(t,\lambda,\nu)-u_n(t,\lambda,\nu)\|\leq$$

$$\leq\left|\varphi_1\left(\int_0^T H_1(s)u_n(s)ds\right)-\varphi_1\left(\int_0^T H_1(s)u_{n-1}(s)ds\right)\right|V_0+$$

$$+\left|\varphi_2\left(\int_0^T H_2(s)u_n(s)ds\right)-\varphi_2\left(\int_0^T H_2(s)u_{n-1}(s)ds\right)\right|V_0+$$

$$+\frac{|\nu|}{\lambda}\sum_{i=1}^{k}\frac{\left|\Delta_{\Psi_{3i}}(f_n,\nu,\lambda)-\Delta_{\Psi_{3i}}(f_{n-1},\nu,\lambda)\right|}{|\Delta_{\Phi}(\nu,\lambda)|}\max_{0\leq t\leq T}|\delta_{13i}(t,\lambda)|+$$

$$+\max_{0\le t\le T}\int_0^T |f(s,u_n(s))-f(s,u_{n-1}(s))|\cdot|\delta_{14}(t,s,\lambda)|\,ds+$$

$$+\max_{0\le t\le T}\int_0^T\int_0^s |f(\theta,u_n(\theta))-f(\theta,u_{n-1}(\theta))|\cdot|\delta_{15}(t,s,\theta,\lambda)|\,d\theta ds\le$$

$$\le\rho\cdot\|u_n(t,\lambda,\nu)-u_{n-1}(t,\lambda,\nu)\|,\tag{4.38}$$

where

$$\rho=L_0\int_0^T[|H_1(s)|+|H_2(s)|]\,ds+(\bar N_1+\bar N_2)\frac{|\nu|}{\lambda}\sum_{i=1}^k\frac{|\bar\Delta_{\Psi_{3i}}(\nu,\lambda)|}{|\Delta_\Phi(\nu,\lambda)|}\max_{0\le t\le T}|\delta_{13i}(t,\lambda)|+$$

$$+\max_{0\le t\le T}\int_0^T|L(s)|\cdot|\delta_{14}(t,s,\lambda)|\,ds$$

$$+\max_{0\le t\le T}\int_0^T\int_0^s|L(\theta)|\cdot|\delta_{15}(t,s,\theta,\lambda)|\,d\theta\,ds<1.\tag{4.39}$$

According to (4.39), $\rho<1$. Consequently, from the estimates (4.36) and (4.38) we deduce that the solution of the nonlinear functional integral equation (4.30) exists and is unique on the interval $[0,T]$.

5. Irregular values of the kernel

On the set Ω_5 of the values of parameters λ and ν, the solution of problem (4.1)–(4.3) is reduced to the following homogeneous system of algebraic equations (HSAE):

$$\tau_i-\frac{\nu}{\lambda}\sum_{j=1}^k\tau_i\Phi_{ij}=0,\quad i=\overline{1,k},\tag{4.40}$$

if the orthogonality condition

$$\begin{cases}\Psi_{1i}(\lambda)=0, & \lambda\in\Lambda_5,\\ \Psi_{2i}(\lambda)=0, & \lambda\in\Lambda_5,\\ \Psi_{3i}(f,\lambda)=0, & \lambda\in\Lambda_5\end{cases}\tag{4.41}$$

is satisfied, where $\Psi_{ji}(\lambda)$ $(j=1,2,3)$ are defined by (4.25)–(4.27). Let us check the fulfillment of condition (4.41) for the values of the parameter λ

5. Irregular values of the kernel

On the set Ω_5 of the values of parameters λ and ν, the solution of problem (4.1)–(4.3) is reduced to the following homogeneous system of algebraic equations (HSAE):

$$\tau_i - \frac{\nu}{\lambda}\sum_{j=1}^{k}\tau_i\Phi_{ij} = 0, \quad i = \overline{1,k}, \tag{4.40}$$

if the orthogonality condition

$$\begin{cases} \Psi_{1i}(\lambda) = 0, \quad \lambda \in \Lambda_5, \\ \Psi_{2i}(\lambda) = 0, \quad \lambda \in \Lambda_5, \\ \Psi_{3i}(f,\lambda) = 0, \quad \lambda \in \Lambda_5 \end{cases} \tag{4.41}$$

is satisfied, where $\Psi_{ji}(\lambda)$ $(j = 1,2,3)$ are defined by (4.25)–(4.27). Let us check the fulfillment of condition (4.41) for the values of the parameter λ from the set Λ_5. The condition (4.41) is reduced to the following system of equations:

$$\begin{cases} \int\limits_0^T [t\cdot\delta_{11}(t,\lambda) - (T-t)\,\delta_{21}(t,\lambda)]\,dt = 0, \quad \lambda \in \Lambda_5, \\ \int\limits_0^T [-t\cdot\delta_{12}(t,\lambda) + (T-t)\,\delta_{22}(t,\lambda)]\,dt = 0, \quad \lambda \in \Lambda_5, \\ \int\limits_0^T [t\cdot\delta_{14}(t,\lambda) + (T-t)\,\delta_{24}(t,\lambda)]\,dt + \\ + \int\limits_0^T\int\limits_0^t [s\cdot\delta_{15}(t,s,\lambda) + (T-s)\,\delta_{25}(t,s,\lambda)]\,ds\,dt = 0, \quad \lambda \in \Lambda_5. \end{cases} \tag{4.42}$$

However, the set of solutions to equation (4.42) cannot intersect with the set of solutions to equation (4.19). Recall that the set of solutions to equation (4.19) is denoted by $\Im_6$. The set Λ_5 denotes $\cup_{i=1}^{4}\Lambda_i \setminus \Im_6$. The system (4.42) is based on the assumption that $\lambda \notin \Im_6$. Therefore, the values of the parameter, for which the orthogonality conditions (4.42) are satisfied, lie in the set Λ_5. So, we compose a new set $\Omega_6 = \{(\lambda,\nu) : \lambda \in \Im_7,\ \nu = \lambda\mu_r\}$, where the set $\Im_7$ denotes the values of the parameter, for which the orthogonality conditions (4.41) are satisfied.

We will construct an infinite set of solutions to problem (4.1)–(4.3) on the set Ω_6. HSAE (4.40) has some number $p\,(1 \le p < k)$ of linearly independent nonzero vector-solutions $\left\{\tau_1^{(l)}, \tau_2^{(l)}, ..., \tau_k^{(l)}\right\}$, $l = \overline{1,p}$. As we know from the theory of Fredholm integral equations, the functions

$$u_l(t,\lambda,\nu) = \frac{\nu}{\lambda}\sum_{i=1}^{k} \tau_i^{(l)} \xi_i(t), \ \ l = \overline{1,p}$$

are nontrivial solutions of the corresponding homogeneous equation

$$u(t,\lambda,\nu) = \frac{\nu}{\lambda}\sum_{i=1}^{k} \xi_i(t) \int_{-T}^{T} b_i(s) u(s,\lambda,\nu) ds, \tag{4.43}$$

where

$$\xi_i(t) = \int_0^t \left[s \cdot \delta_{13i}(s,\lambda) - (T-s)\,\delta_{23i}(s,\lambda)\right] ds.$$

Therefore, the general solution of the homogeneous integral equation (4.43) can be written as

$$u(t,\lambda,\nu) = \sum_{l=1}^{p} \alpha_l u_l(t,\lambda,\nu), \tag{4.44}$$

where α_l are arbitrary constants. One can easily see that the solutions (4.44) do not depend on nonlinear functions.

It is obvious that on the set $\Omega_7 = \Omega_5 \setminus \Omega_6 = \left\{(\lambda,\nu) : \lambda \in \Lambda_5 \setminus \Im_7, \ \nu = \lambda\mu_r\right\}$, the orthogonality condition (4.41) is not satisfied. Consequently, the problem (4.1)–(4.3) has no solution.

6. Irregular values of parameter λ

Now consider irregular values of parameter $\lambda \in \Im_6$. In this case, the main determinant of the transcendental system of algebraic equations (4.7) vanishes: $D(\lambda) = \chi_{11}(\lambda)\chi_{22}(\lambda) - \chi_{12}(\lambda)\chi_{21}(\lambda) = 0$. Then we cannot uniquely solve the linear system of equations (4.7). So, we cannot uniquely determine the coefficients A_1 and A_2 Therefore, the problem (4.1)–(4.3) can have an infinite set of solutions. We will study the sufficient conditions for the existence of an infinite set of solutions. Substituting representations (4.5) and (4.6) into (4.4), we arrive at a new system of linear equations (SLE):

$$\tau_i - \frac{\nu}{\lambda}\sum_{j=1}^{k} \tau_j \Phi_{ij}(\lambda) = A_1 Q_{1i}(\lambda) + A_2 Q_{2i}(\lambda) + Q_{2i}(f,\lambda), \ \ i = \overline{1,k}, \tag{4.45}$$

where

$$\Phi_{ij}(\lambda) = \int_0^T b_i(s) \left[\int_0^s a_j(\theta) \left[s \cdot \sin\lambda\,(s-\theta) + \lambda(T-s)\cos\lambda\,(s-\theta) \right] d\theta \right] ds,$$

$$Q_{1i}(\lambda) = \int_0^T b_i(s) \left[s \cdot \cos\lambda s - \lambda(T-s)\sin\lambda s \right] ds, \tag{4.46}$$

$$Q_{2i}(\lambda) = \int_0^T b_i(s) \left[s \cdot \sin\lambda s + \lambda(T-s)\cos\lambda s \right] ds, \tag{4.47}$$

$$Q_{3i}(f,\lambda) = \int_0^T b_i(s) \int_0^s f(\theta,u(\theta)) \left[\frac{s}{\lambda}\sin\lambda(s-\theta) + (T-s)\cos\lambda(s-\theta) \right] d\theta ds, \tag{4.48}$$

A_1, A_2 are arbitrary constants. It is known that SLE (4.45) is uniquely solvable for any finite right-hand sides Q_i, if the Fredholm condition (4.28) is satisfied.

Note that the unique solvability of SLE (4.45) does not guarantee the uniqueness of the solution to problem (4.1)–(4.3). We construct only an infinite set of solutions to problem (4.1)–(4.3) for the values $\lambda \in \Im_6$. The determinant $\Delta_\Phi(\nu,\lambda)$ in (4.28) is a polynomial with respect to $\dfrac{\nu}{\lambda}$ of degree at most k. The equation has no more than k different real roots. They were denoted above by μ_r, $1 \le r \le k$. The values $\nu = \lambda\mu_r$ were called above the eigenvalues of the kernel of the Fredholm integro-differential equation (4.1) or irregular values of the parameter ν. Other values $\nu \neq \lambda\mu_r$ were called the regular values of the parameter ν, and condition (4.28) is satisfied for them, i.e. $\Delta_\Phi(\nu,\lambda) \neq 0$.

Consider the following two sets:

$$\Omega_{8,1} = \left\{ (\lambda,\nu) : \lambda \in \Im_6,\ \nu \neq \lambda\,\mu_r \right\}, \quad \Omega_{8,2} = \left\{ (\lambda,\nu) : \lambda \in \Im_6,\ \nu = \lambda\,\mu_r \right\}.$$

On the set $(\lambda,\nu) \in \Omega_{8,1}$ the solution of SAE (4.45) is written as

$$\tau_i = A_1 \frac{\Delta_{Q_{1i}}(\lambda,\nu)}{\Delta_\Phi(\lambda,\nu)} + A_2 \frac{\Delta_{Q_{2i}}(\lambda,\nu)}{\Delta_\Phi(\lambda,\nu)} + \frac{\Delta_{Q_{3i}}(f,\lambda,\nu)}{\Delta_\Phi(\lambda,\nu)}, \quad i = \overline{1,k}, \tag{4.49}$$

where

$$\Delta_{Q_{mi}}(\lambda,\nu) = \begin{vmatrix} 1-\frac{\nu}{\lambda}\Phi_{11} & \cdots & \frac{\nu}{\lambda}\Phi_{1(i-1)} & Q_{m1} & \frac{\nu}{\lambda}\Phi_{1(i+1)} & \cdots & \frac{\nu}{\lambda}\Phi_{1k} \\ \frac{\nu}{\lambda}\Phi_{21} & \cdots & \frac{\nu}{\lambda}\Phi_{2(i-1)} & Q_{m2} & \frac{\nu}{\lambda}\Phi_{2(i+1)} & \cdots & \frac{\nu}{\lambda}\Phi_{2k} \\ \cdots & \cdots & \cdots & \cdots & \cdots & \cdots & \cdots \\ \frac{\nu}{\lambda}\Phi_{k1} & \cdots & \frac{\nu}{\lambda}\Phi_{k(i-1)} & Q_{mk} & \frac{\nu}{\lambda}\Phi_{k(i+1)} & \cdots & 1-\frac{\nu}{\lambda}\Phi_{kk} \end{vmatrix},$$

$m = 1, 2, 3$.

Substituting (4.49) into (4.5), we obtain an infinite set of solutions to problem (4.1)–(4.3) on the set $(\lambda, \nu) \in \Omega_{8,1}$:

$$u(t, \lambda, \nu) = A_1 P_1(t, \lambda, \nu) + A_2 P_2(t, \lambda, \nu) +$$

$$+\frac{\nu}{\lambda}\sum_{i=1}^{k}\frac{\Delta_{Q_{3i}}(f, \lambda, \nu)}{\Delta_\Phi(\lambda, \nu)}\int_0^t \sin\lambda(t-s)a_i(s)ds + \frac{1}{\lambda}\int_0^1 \sin\lambda(t-s)f(s, u(s))ds, \tag{4.50}$$

where A_1 and A_2 are arbitrary constants, and

$$P_1(t, \lambda, \nu) = \cos\lambda t + \frac{\nu}{\lambda}\sum_{i=1}^{k}\frac{\Delta_{Q_{1i}}(\lambda, \nu)}{\Delta_\Phi(\lambda, \nu)}\int_0^t \sin\lambda\,(t-s)\,a_i(s)ds,$$

$$P_2(t, \lambda, \nu) = \sin\lambda t + \frac{\nu}{\lambda}\sum_{i=1}^{k}\frac{\Delta_{Q_{2i}}(\lambda, \nu)}{\Delta_\Phi(\lambda, \nu)}\int_0^t \sin\lambda\,(t-s)\,a_i(s)ds.$$

Proof of existence of solutions to the nonlinear functional integral equation (4.50) is similar to corresponding proof for the nonlinear functional integral equation (4.30).

On the set $(\lambda, \nu) \in \Omega_{8,2}$, the problem (4.1)–(4.3) reduces to the HSAE (4.40), if the orthogonality conditions

$$\begin{cases} Q_{1i}(\lambda) = 0, \\ Q_{2i}(\lambda) = 0, \qquad \lambda \in \Im_6 \\ Q_{3i}(f, \lambda) = 0, \end{cases} \tag{4.51}$$

are satisfied, where $Q_{mi}(\lambda)\,(m = 1, 2, 3)$ are defined by (4.46), (4.47) and (4.48), respectively. The condition (4.51) reduces to the following three equations as a system for the set $\Im_6$:

$$\cos(\lambda T - \theta) - \frac{1+\lambda^2 T}{\sqrt{1+(1+T)^2\lambda^2}} = 0, \ \theta = \arccos\frac{1}{\sqrt{1+(1+T)^2\lambda^2}}, \tag{4.52}$$

$$\cos(\lambda T + \xi) - \frac{\lambda}{\sqrt{1+(1+T)^2\lambda^2}} = 0, \ \xi = \arccos\frac{(1+T)\lambda}{\sqrt{1+(1+T)^2\lambda^2}}, \tag{4.53}$$

$$\cos\lambda T - \lambda T = 0. \tag{4.54}$$

The graphs of the functions in (4.52), (4.53) and (4.54) are built in the coordinate system at $T = 2$ (see, Figure 8). As we see from the graph, the system of equations (4.51) has no solution for the values of the parameter $\lambda \in \Im_6$. Therefore, HSAE (4.40) is meaningless in this case. Consequently, problem (4.1)–(4.3) has no solution on the set $(\nu, \lambda) \in \Omega_{8,2}$.

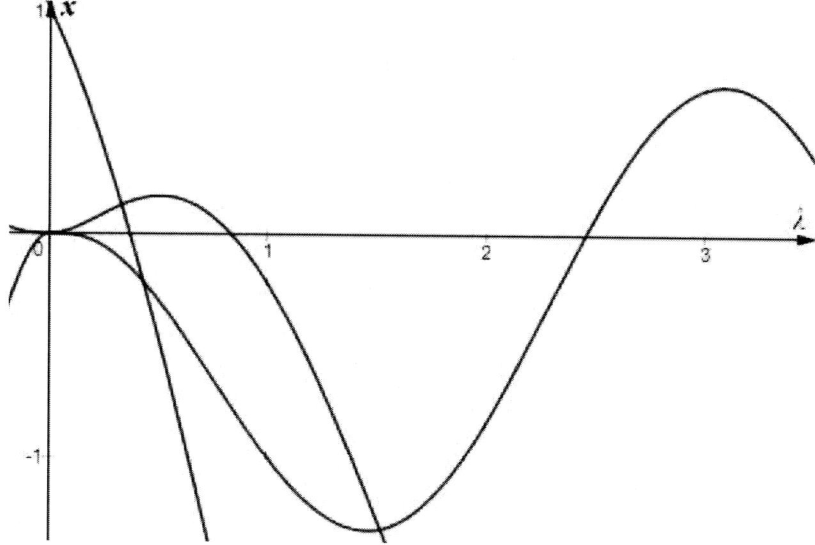

Figure 4.8. Graphs of the functions (4.44), (4.45) and (4.46). T=2

7. Conclusion

In this chapter, we explored the solvability and solution construction for the second-order nonlinear Fredholm integro-differential equation (4.1) with nonlinear integral conditions (4.2) and (4.3). The degenerate kernel method was further developed, and specific challenges in constructing solutions, particularly those related to determining integration coefficients, were addressed. The parameter values λ and ν for which the nonlocal problem is solvable were identified, and the corresponding solutions were derived. The following theorem has been proved.

In this chapter, we considered the issues of solvability and construction of solutions for the second order nonlinear Fredholm integro-differential equation (4.1) with nonlinear integral conditions (4.2) and (4.3). The degenerate kernel method was developed. We also studied some circumstances that had arisen in the construction of solutions. These circumstances are associated with the determination of the integration coefficients. The values of the parameters λ and ν are calculated, for which the solvability of the nonlocal problem is established and the corresponding solutions are constructed.

The following theorem has been proved.

Theorem 7.1. *The following assertions for the solvability of the nonlinear* *problem* (4.1)–(4.3) *are true.*

1. For the values of the parameter λ, *for which* $\chi_{11}(\lambda) = \chi_{12}(\lambda) = 0$, *the nonlinear problem* (4.1)–(4.3) *has no solution, if the condition* (4.14) *is fulfilled. On the set* $\Omega_7 = \{(\lambda, \nu) : \lambda \in \Lambda_5 \setminus \Im_7,\ \nu = \lambda \mu_r\}$ *the nonlinear problem* (4.1)–(4.3) *has no solution. The nonlinear problem* (4.1)–(4.3) *has no solution on the set* $(\nu, \lambda) \in \Omega_{8,2} = \{(\lambda, \nu) : \lambda \in \Im_6,\ \nu = \lambda \mu_r\}$, *too.*

2. On the set $\tilde{\Omega}_5 = \{\lambda \in \Lambda_5,\ \nu \neq \lambda \mu_r\}$ *of parameters* (ν, λ), *the nonlinear problem* (4.1)–(4.3) *has a unique solution, if the conditions* (4.31)–(4.34) *and* (4.39) *hold. This unique solution can be found from the iteration process* (4.35).

3. On the set $\Omega_6 = \{(\lambda, \nu) : \lambda \in \Im_7,\ \nu = \lambda \mu_r\}$ *of parameters* (ν, λ), *the nonlinear problem* (4.1)–(4.3) *has an infinite set of solutions, and these solutions are presented by formula* (4.44). *On the set* $\Omega_{8,1} = \{(\lambda, \nu) : \lambda \in \Im_6,\ \nu \neq \lambda \mu_r\}$ *of parameters* (ν, λ), *the nonlinear problem* (4.1)–(4.3) *has an infinite set of solutions, if the conditions* (4.31) *and* (4.32) *are fulfilled. These solutions are presented as nonlinear functional integral equation* (4.50). *The equation* (4.50) *can be solved by the method of successive approximations.*

Chapter 5

On the Approximate Solution of the Cauchy Problem for Elliptic Systems of the First Order

D.A. Juraev *
Department of Scientific Research, Innovation and Training of Scientific and Pedagogical Staff,
University of Economics and Pedagogy, Karshi 180100, Uzbekistan
Department of Mathematics, Anand International College of Engineering, Jaipur 303012, India

Abstract

In this chapter, we discuss the formulation of the Cauchy problem for matrix factorizations of the Helmholtz equation in a four-dimensional bounded domain. Preliminary information and the formulation of the Cauchy problem are provided. Based on the constructed Carleman function, a regularized solution of the Cauchy problem for the matrix factorization of the Helmholtz equation in a four-dimensional bounded domain is presented in explicit form.

This study aims to provide an understanding of well-posed and ill-posed problems, as well as the methods developed for solving ill-posed applied problems in mathematics. The history and significance of ill-posed problems in addressing various applied issues in the natural sciences are explored in detail. The study of methods for solving ill-posed problems has garnered significant interest among researchers, who are actively conducting research in this field. The

*Corresponding Author's Email: juraevdavron12@gmail.com

theory of ill-posed problems is a rapidly developing area in mathematical physics and the natural sciences. In practice, most problems are ill-posed, requiring decision-making under conditions of uncertainty, overdetermination, or inconsistency. The main conclusion drawn from this study is that solving ill-posed problems cannot be accomplished solely by learning from well-posed problems.

Keywords: Ill-posed problems, Carleman functions, Helmholtz equation, approximate solutions, Cauchy problem.

AMS Subject Classification: 35J46, 35J56.

1. Introduction

The ubiquity of ill-posed problems is rooted in their occurrence across various scientific disciplines, from physics to finance, where models often rely on incomplete, noisy, or insufficiently precise data. This inherent uncertainty challenges traditional frameworks of problem-solving, necessitating a deeper understanding of stability and solution behavior. The emergence of computational techniques has bolstered interest in these problems, enabling innovative approaches that reconcile theoretical principles with real-world applications. Furthermore, the development of regularization methods has become central to addressing ill-posed problems. These techniques introduce additional constraints or information, transforming problematic situations into those amenable to analysis. This interplay between rigor and adaptability has paved the way for advancements in fields such as image reconstruction, signal processing, and machine learning, where data can be sparse or contaminated. The interdisciplinary nature of ill-posed problems fosters collaboration across various domains, encouraging mathematicians, scientists, and engineers to share insights and methodologies. As scholars continue to explore these seemingly trivial issues, the sophistication of the underlying theories expands, revealing a rich tapestry of connections that were previously overlooked. Ultimately, the evolution of ill-posed problems reflects both the complexity of the real world and the ingenuity required to navigate it.

Numerous applied issues, including those in geophysics, biophysics, electrodynamics, gas dynamics, and plasma physics, can be formulated using the equations of mathematical physics. The creation of a mathematical physics equation that accurately reflects specific physical laws in our environment can be viewed as addressing what may be recognized as an inverse problem. The investigator examines a particular phenomenon and seeks to formulate an equation whose solution exhibits the observed characteristics. The equations typically stem from established physical laws, which allow for the development of a general framework of differential relationships. These equations often incorporate several arbitrary functions that define the characteristics of the physical medium. When the medium's properties are known, the mathematical physics equation, when combined with boundary and initial conditions, facilitates predictions regarding the evolution of a physical phenomenon in a given space-time context. This formulation represents a classical scenario in mathematical physics. In the realm of inverse problems, these instances are termed direct problems. While contemporary natural science is familiar with the general format of these equations, the specific traits of the medium need to be extracted from the observed solutions of the equations.

The concept of the correctness of a problem statement in mathematical physics was formulated at the beginning of the 20th century by the famous French mathematician J. Hadamard [207]. The necessity to solve nonstationary problems similar to the one presented above, requires a more exact determination of problem's solution. In conditionally well-posed problems, according to Tikhonov, we are not just dealing with a solution, but with a solution that belongs to some class of solutions. Narrowing the class of admissible solutions allows, in some cases, for the problem to become well-posed.

We say that a problem is Tikhonov well-posed if:

1) the solution of the problem is a priori known to exist in some class;

2) in this class, the solution is unique;

3) the solution continuously depends on input data.

The fundamental difference here consists of separating out the class of admissible solutions. The classes of a priori restrictions differ widely. In the consideration of ill-posed problems, the problem statement itself undergoes substantial changes: the condition that the solution belongs to a certain set must be included in the problem statement. J. Hadamard illustrated this with the example of the Cauchy problem for the Laplace equation, which has become a classic example of an ill-posed problem. The need to consider problems of mathematical physics that are incorrect in the classical sense (according to Hadamard) in connection with the interpretation of geophysical observational data was first indicated in 1943 by A.N. Tikhonov [248]. He showed that if the class of possible solutions is reduced to a compact set, then the existence and uniqueness of the solution implies its stability. Ways of development of the theory and methods for solving ill-posed problems are associated with the names of prominent mathematicians A.N. Tikhonov, M.M. Lavrent'ev, V.K. Ivanov, as well as with the mathematical schools they created, which largely determined the development of the theories and applications of ill-posed problems.

A large number of problems in mathematical physics that do not satisfy the Hadamard correctness conditions can be reduced to an operator equation of the first kind. Since the problems of mathematical physics describe real processes in nature, they must satisfy certain requirements. The stability requirement means that any physically defined process must continuously depend on the initial and boundary conditions and on the inhomogeneous term in the equation; that is, it should be characterized by functions that change little with small changes in the initial data. Processes that do not meet this criterion are not physically defined. Stability is also important for the approximate solution of problems.

Among mathematical problems, a class of problems stands out, the solutions of which are unstable to small changes in the initial data. They are characterized by the fact that arbitrarily small changes in the initial data can lead to arbitrarily large changes in the solutions. Problems of this type are, in essence, ill-posed and belong to the category of ill-posed problems.

In the last three decades, regularization methods have been proposed for solving ill-posed problems. However, deterministic methods have been significantly used to introduce a priori information about both the solution itself and the errors in the initial data of the problem. Unreasonably little attention is paid to the choice of optimal values for the parameters of algorithms, which would allow for obtaining solutions with the smallest error, as well as for the construction of algorithms with given accuracy characteristics. There are no effective algorithms that take into account the available a priori information about the desired solution (for example, regarding the range of possible values of the coefficients of the identified model). The lack of software developed in the environment of a universal mathematical package (for example, Mathcad) creates significant difficulties for engineers and experimenters (who are not programmers) in using regularizing algorithms in practice. Tasks that do not satisfy all of the above requirements 1) - 3) are, according to Hadamard, incorrectly delivered. In 1926, T. Carleman (see, for instance [202], p. 41) constructed a formula that connects the values of the analytic function of a complex variable at the points of the region with its values on a piece of the boundary of this region. The construction of the Carleman function allows for the regularization of these problems and provides an estimate of conditional stability. It is known that the Helmholtz equation has fundamentally different solutions in various spaces. Using the construction of a fundamental solution, we will develop an approximate solution for the Helmholtz equation. M.M. Lavrent'ev, in his works on the Cauchy problem for the Laplace equation and other ill-posed problems in mathematical physics, proposed a method for distinguishing the correctness class and developed stable methods for solving them (see, for instance, [241]- [242]). He also introduced the construction of a regularized solution for the Cauchy problem of the Laplace equation using the Carleman function. Additionally, in the 1970s, Sh. Yarmukhamedov developed a method for constructing a family of fundamental solutions parametrized by an entire function with certain properties [249]. This method is used to construct explicit formulas that restore solutions of elliptic equations in a domain from their Cauchy data on a piece of the domain boundary. Such formulas are also called Carleman formulas. The multidimensional Carleman formula was constructed by L.A. Aizenberg [196]. In unstable problems, the image of the operator is not closed, therefore, the solvability condition cannot be written in terms of continuous linear functionals. So, in the Cauchy problem for elliptic equations with data on a part of the boundary of a domain, the solution is usually unique, the problem is solvable for an everywhere dense data set, but this set is not closed. Consequently, the theory of solvability of such problems is much more difficult and deeper than the theory of solvability of the Fredholm equations. The first results in this direction appeared only in the mid-1980s in the works of L.A. Aizenberg, A.M. Kytmanov and N.N. Tarkhanov [247]. An analogue of the Carleman formula for one class of elliptic systems with constant coefficients on the plane is considered in the work of E.V. Arbuzov and A.L. Bukhgeim [197]. The construction of the Carleman matrix for elliptic systems was carried out by Sh. Yarmukhamedov, N.N. Tarkhanov, I.E. Niyozov and others. In [200]- [225], the exact and approximate solutions of the ill-posed Cauchy problem for various factorizations of the Helmholtz equation are studied. Such problems arise in mathematical physics and in various fields of natural science (for example, in electro-geological exploration, cardiology, electrodynamics, etc.). Using the methods of previous works, the validity of the fundamental solution for the matrix factorization of the Helmholtz equation in various spaces was proved in [213]- [238] and [251]. Currently, the theory of ill-posed problems is one of the pressing issues in the field of partial differential equations. Many scientific and applied problems, studied at the world level, are in many cases reduced to the ill-posed boundary value problems for partial differential equations. Applied research on conditional correctness and construction of an approximate solution for given values on a part of the boundary of the region, for equations of elliptical type, are especially important in hydrodynamics, geophysics and electrodynamics. The study of a family of regularizing solutions to ill-posed problems served as an impetus for the beginning of studies of the well-posedness class when narrowed to a compact set. Therefore, the study of ill-posed problems for linear elliptic systems of the first order is one of the topical problems in the theory of partial differential equations. At present, in the world, in the study of ill-posed boundary value problems for linear elliptic systems of the first order, the construction of a regularized solution plays a special role. The Cauchy problem for elliptic equations is ill-posed (see, for instance, [207], p. 39). Boundary problems, as well as numerical solutions of some problems, are considered in [198]- [201], [204]- [205], [208]- [212], [239, 243, 244] and [247]- [250].

At present, special attention is paid to topical aspects of differential equations and mathematical physics, which have scientific and practical applications in the fundamental sciences. In particular, there is a focus on the study of various ill-posed boundary value problems for partial differential equations of elliptic type, which have practical applications in applied sciences. As a result, significant findings have been obtained in the study of ill-posed boundary value problems for partial differential equations. Specifically, approximate solutions were constructed using Carleman matrices in explicit form from approximate data in special domains, and estimates of conditional stability and solvability criteria were established. The first results of practical importance for ill-posed problems, along with efforts to reduce the class of possible solutions to a compact set and to transform problems into stable ones, were achieved by A.N. Tikhonov (see [248]). M.M. Lavrent'ev obtained the estimates that characterize the stability of the spatial problem in the class of bounded solutions of the Cauchy problem for the Laplace equation and some other ill-posed problems of mathematical physics in a straight cylinder, as well as for an arbitrary spatial domain with a sufficiently smooth boundary (see, for instance, [241]- [242]).

In this chapter, based on the results of [203, 206], [236], [241, 242], [245]- [250] and the Cauchy problem for the Laplace and Helmholtz equations, an explicit Carleman matrix is constructed and, on its basis, a regularized solution of the Cauchy problem for the matrix factorization of the Helmholtz equation is obtained.

The problem of reconstructing the solution for matrix factorization of the Helmholtz equation (see, for instance, [213]- [239] and [251]), is one of the topical problems in the theory of differential equations.

Today, there is still interest in classical ill-posed problems of mathematical physics. This direction in the study of the properties of solutions of Cauchy problem for Laplace equation dates back to [202, 203], [206], [241]- [242], [249]- [250] and was subsequently developed in [197, 245–247], [213]- [239] and [251].

Let $\mathbb{R}^4$ be a four-dimensional real Euclidean space,

$$\zeta = (\zeta_1, \zeta_2, \zeta_3, \zeta_4) \in \mathbb{R}^4, \quad \eta = (\eta_1, \eta_2, \eta_3, \eta_4) \in \mathbb{R}^4,$$
$$\zeta' = (\zeta_1, \zeta_2, \zeta_3) \in \mathbb{R}^3 \quad \eta' = (\eta_1, \eta_2, \eta_3) \in \mathbb{R}^3.$$

We introduce the following notation:

$$r = |\eta - \zeta|, \quad \alpha = |\eta' - \zeta'|, \quad z = i\sqrt{a^2 + \alpha^2} + \eta_4, \quad a \geq 0,$$

$$\partial_\zeta = \left(\partial_{\zeta_1}, \partial_{\zeta_2}, \partial_{\zeta_3}, \partial_{\zeta_4}\right)^T, \quad \partial_\zeta = \chi^T, \quad \chi^T = \begin{pmatrix} \chi_1 \\ \chi_1 \\ \chi_3 \\ \chi_4 \end{pmatrix} \text{ is a transposed vector}$$

χ,

$$\mathfrak{W}(\zeta) = (\mathfrak{W}_1(\zeta), \ldots, \mathfrak{W}_n(\zeta))^T, \quad v^0 = (1, \ldots, 1) \in \mathbb{R}^n, \quad n = 2^4,$$

$$E(w) = \begin{Vmatrix} w_1 & 0 & \cdots & 0 \\ 0 & w_2 & \cdots & 0 \\ \cdots & \cdots & \ddots & \cdots \\ 0 & 0 & 0 & w_n \end{Vmatrix} \text{isadiagonal matrix}, w = (w_1, \ldots, w_n) \in \mathbb{R}^n.$$

We also consider a bounded simply-connected domain $\Omega \subset \mathbb{R}^4$ with a piecewise smooth boundary $\partial\Omega = \Sigma \bigcup D$, where Σ is a smooth surface lying in the half-space $\eta_m > 0$ and D is the plane $\eta_4 = 0$.

$\mathfrak{P}(\chi^T)$ is an $(n \times n)-$dimensional matrix satisfying

$$\mathfrak{P}^*(\chi^T)\mathfrak{P}(\chi^T) = E((|\chi|^2 + \lambda^2)v^0),$$

where $\mathfrak{P}^*(\chi^T)$ is the Hermitian conjugate matrix of $\mathfrak{P}(\chi^T)$, $\lambda \in \mathbb{R}$, the elements of the matrix $\mathfrak{P}(\chi^T)$ consist of a set of linear functions with constant coefficients from the complex plane $\mathbb{C}$.

Let us consider the following first order systems of linear partial differential equations with constant coefficients

$$\mathfrak{P}\left(\partial_\zeta\right)\mathfrak{W}(\zeta) = 0 \tag{5.1}$$

in the domain Ω, where $P\left(\partial_\zeta\right)$ is a matrix differential operator of the first-order.

Also consider the set

$$S(\Omega) = \left\{\mathfrak{W} : \overline{\Omega} \longrightarrow \mathbb{R}^n \mid \mathfrak{W} \text{ is continuous on } \overline{\Omega} = \Omega \cup \partial\Omega \text{ and } \mathfrak{W} \text{ satisfies the system (5.1)}\right\}.$$

2. Statement of the Cauchy problem

Formulation of the problem. Suppose $\mathfrak{W}(\eta) \in S(\Omega)$ and

$$\mathfrak{W}(\eta)|_\Sigma = f(\eta), \quad \eta \in \Sigma. \tag{5.2}$$

Here, $f(\eta)$ is given continuous vector-function on Σ. It is required to restore the vector function $W(\eta)$ in the domain Ω, based on its values $f(\eta)$ on Σ.

If $\mathfrak{W}(\eta) \in S(\Omega)$, then the following integral formula of Cauchy type is valid:

$$\mathfrak{W}(\zeta) = \int\limits_{\partial\Omega} \mathfrak{L}(\eta, \zeta; \lambda) \mathfrak{W}(\eta) ds_\eta, \quad \zeta \in \Omega, \tag{5.3}$$

where

$$\mathfrak{L}(\eta, \zeta; \lambda) = \left(E\left(\Gamma_m(\lambda r) v^0\right) \mathfrak{P}^*\left(\partial_\zeta\right)\right) \mathfrak{P}(t^T).$$

Here $t = (t_1, t_2, t_3, t_4)$ is an outer unit normal drawn at the point η, $\partial\Omega$ is a surface, and $\Gamma_4(\lambda r)$ is a fundamental solution of the Helmholtz equation in $\mathbb{R}^4$, where $\Gamma_4(\lambda r)$ is defined by the following formula:

$$\begin{gathered} \Gamma_4(\lambda r) = B_4 \lambda \frac{H_1^{(1)}(\lambda r)}{r}, \\ B_4 = \frac{1}{4i\pi}. \end{gathered} \tag{5.4}$$

Here $H_1^{(1)}(\lambda r)$ is a $1-$th order the Hankel function of the first kind (see, for instance, [240]).

Let $K(z)$ be an entire function taking real values for real z, $(z = a + ib,\ a, b \in \mathbb{R})$ such that

$$\begin{gathered} K(a) \neq 0, \quad \sup_{b \geq 1} \left| b^p K^{(p)}(z) \right| = N(a, p) < \infty, \\ -\infty < a < \infty, \quad p = \overline{0, 4}. \end{gathered} \tag{5.5}$$

We define the function $\Psi(\eta, \zeta; \lambda)$ at $\eta \neq \zeta$ by the following equality:

$$\Psi(\eta, \zeta; \lambda) = \frac{1}{c_4 K(\zeta_4)} \frac{\partial}{\partial s} \int\limits_0^\infty \operatorname{Im}\left[\frac{K(z)}{z - \zeta_4}\right] \frac{a I_0(\lambda a)}{\sqrt{a^2 + \alpha^2}} da, \tag{5.6}$$

where $c_4 = 2\omega_4$; $I_0(\lambda a) = J_0(i\lambda a)$ is a zero order Bessel function of the first kind [240], and ω_4 is an area of a unit sphere in the space $\mathbb{R}^4$.

In the formula (5.6), choosing

$$K(z) = \exp(\sigma z), \quad K(\zeta_4) = \exp(\sigma \zeta_4), \quad \sigma > 0, \tag{5.7}$$

we get

$$\Psi_\sigma(\eta, \zeta; \lambda) = \frac{e^{-\sigma\zeta_4}}{c_4} \frac{\partial}{\partial s} \int\limits_0^\infty \operatorname{Im}\left[\frac{\exp(\sigma z)}{z - \zeta_4}\right] \frac{a I_0(\lambda a)}{\sqrt{a^2 + \alpha^2}} da. \tag{5.8}$$

The formula (5.3) is true if we substitute $\Gamma_m(\lambda r)$ with the function

$$\Psi_\sigma(\eta,\zeta;\lambda)=\Gamma_4(\lambda r)+G_\sigma(\eta,\zeta;\lambda), \tag{5.9}$$

where $G_\sigma(\eta,\zeta;\lambda)$ is a regular solution of the Helmholtz equation with respect to the variable η, including the point $\eta=\zeta$.

Then the integral formula has the form

$$\mathfrak{W}(\zeta)=\int\limits_{\partial\Omega}\mathfrak{L}_\sigma(\eta,\zeta;\lambda)\mathfrak{W}(\eta)ds_\eta, \quad \zeta\in\Omega, \tag{5.10}$$

where

$$\mathfrak{L}_\sigma(\eta,\zeta;\lambda)=\left(E\left(\Psi_\sigma(\eta,\zeta;\lambda)v^0\right)\mathfrak{P}^*\left(\partial_\zeta\right)\right)\mathfrak{P}(t^T).$$

3. Solution of the Cauchy problem (5.1)-(5.2)

Theorem 3.1. *Let $\mathfrak{W}(\eta)\in S(\Omega)$ satisfy the inequality*

$$|\mathfrak{W}(\eta)|\le M, \quad \eta\in D. \tag{5.11}$$

If

$$\mathfrak{W}_\sigma(\zeta)=\int\limits_{\Sigma}\mathfrak{L}_\sigma(\eta,\zeta;\lambda)\mathfrak{W}(\eta)ds_y, \quad \zeta\in\Omega, \tag{5.12}$$

then the following estimates are true:

$$|\mathfrak{W}(\zeta)-\mathfrak{W}_\sigma(\zeta)|\le M\mathfrak{F}(\lambda,\zeta)\sigma^2e^{-\sigma\zeta_4}, \quad \sigma>1, \quad \zeta\in\Omega. \tag{5.13}$$

$$\left|\frac{\partial\mathfrak{W}(\zeta)}{\partial\zeta_j}-\frac{\partial\mathfrak{W}_\sigma(\zeta)}{\partial\zeta_j}\right|\le M\mathfrak{F}(\lambda,\zeta)\sigma^2e^{-\sigma\zeta_4}, \quad \sigma>1, \quad \zeta\in\Omega, \quad j=\overline{1,4}. \tag{5.14}$$

Here and below, the functions bounded on compact subsets of the domain Ω are denoted by $\mathfrak{F}(\lambda,\zeta)$.

Proof. Let us first estimate the inequality (5.13). Using the integral formula (5.10) and the equality (5.12), we obtain

$$\mathfrak{W}(\zeta)=\int\limits_{\Sigma}\mathfrak{L}_\sigma(\eta,\zeta;\lambda)\mathfrak{W}(\eta)ds_\eta+\int\limits_{D}\mathfrak{L}_\sigma(\eta,\zeta;\lambda)\mathfrak{W}(\eta)ds_\eta=$$

$$=\mathfrak{W}_\sigma(\zeta)+\int\limits_{D}\mathfrak{L}_\sigma(\eta,\zeta;\lambda)\mathfrak{W}(\eta)ds_\eta, \quad \zeta\in\Omega.$$

Taking into account the inequality (5.11), we estimate the following

$$|\mathfrak{W}(\zeta)-\mathfrak{W}_\sigma(\zeta)| \le \left|\int\limits_D \mathfrak{L}_\sigma(\eta,\zeta;\lambda)\mathfrak{W}(\eta)ds_\eta\right| \le$$
$$\le \int\limits_D |\mathfrak{L}_\sigma(\eta,\zeta;\lambda)||\mathfrak{W}(\eta)|ds_\eta \le M\int\limits_D |\mathfrak{L}_\sigma(\eta,\zeta;\lambda)|ds_\eta, \quad \zeta\in\Omega. \tag{5.15}$$

Now let's estimate the integrals $\int\limits_D |\Psi_\sigma(\eta,\zeta;\lambda)|ds_\eta$, $\int\limits_D \left|\frac{\partial\Psi_\sigma(\eta,\zeta;\lambda)}{\partial\eta_j}\right|ds_\eta$, $(j=\overline{1,3})$ and $\int\limits_D \left|\frac{\partial\Psi_\sigma(\eta,\zeta;\lambda)}{\partial\eta_4}\right|ds_\eta$ on the part D of the plane $\eta_4=0$.

Separating the imaginary part of (5.8), we obtain

$$\Psi_\sigma(\eta,\zeta;\lambda)=\frac{e^{\sigma(\eta_4-\zeta_4)}}{c_m}\left[\frac{\partial}{\partial s}\int\limits_0^\infty \frac{\cos\sigma\sqrt{a^2+\alpha^2}}{a^2+r^2}\,aI_0(\lambda a)\,da-\right.$$
$$\left.-\frac{\partial}{\partial s}\int\limits_0^\infty \frac{(\eta_4-\zeta_4)\sin\sigma\sqrt{a^2+\alpha^2}}{a^2+r^2}\frac{aI_0(\lambda a)}{\sqrt{a^2+\alpha^2}}da\right], \quad \zeta_4>0. \tag{5.16}$$

From (5.16) and the inequality

$$I_0(\lambda a)\le\sqrt{\frac{2}{\lambda\pi a}}, \tag{5.17}$$

we have

$$\int\limits_D |\Psi_\sigma(\eta,\zeta;\lambda)|ds_\eta \le \mathfrak{F}(\lambda,\zeta)\sigma^2 e^{-\sigma\zeta_4}, \quad \sigma>1, \quad x\in G, \tag{5.18}$$

To estimate the second integral, we use the equality

$$\frac{\partial\Psi_\sigma(\eta,\zeta;\lambda)}{\partial\eta_j}=\frac{\partial\Psi_\sigma(\eta,\zeta;\lambda)}{\partial s}\frac{\partial s}{\partial\eta_j}=2(\eta_j-\zeta_j)\frac{\partial\Psi_\sigma(\eta,\zeta;\lambda)}{\partial s},$$
$$s=\alpha^2, \quad j=\overline{1,3}. \tag{5.19}$$

Considering equality (5.16), inequality (5.17) and equality (5.19), we obtain

$$\int_D \left| \frac{\partial \Psi_\sigma(\eta,\zeta;\lambda)}{\partial \eta_j} \right| ds_y \le \mathfrak{F}(\lambda,\zeta)\sigma^2 e^{-\sigma\zeta_4}, \quad \sigma > 1, \quad \zeta \in \Omega, \\ j = \overline{1,3}. \tag{5.20}$$

Now we estimate the integral $\int_D \left| \frac{\partial \Psi_\sigma(\eta,\zeta;\lambda)}{\partial \eta_4} \right| ds_\eta$.

Taking into account equality (5.16) and inequality (5.17), we obtain

$$\int_D \left| \frac{\partial \Psi_\sigma(\eta,\zeta;\lambda)}{\partial \eta_4} \right| ds_\eta \le \mathfrak{F}(\lambda,\eta)\sigma^2 e^{-\sigma\zeta_4}, \quad \sigma > 1, \quad \zeta \in \Omega. \tag{5.21}$$

From inequalities (5.18), (5.20) and (5.21), bearing in mind (5.15), we get the estimate (5.13).

Now let us prove inequality (5.14). To do this, we take the derivatives from equalities (5.10) and (5.12) with respect to ζ_j, $j = \overline{1,4}$. Then we obtain the following:

$$\frac{\partial \mathfrak{W}(\zeta)}{\partial \zeta_j} = \int_\Sigma \frac{\partial \mathfrak{L}_\sigma(\eta,\zeta;\lambda)}{\partial \zeta_j} \mathfrak{W}(\eta) ds_\eta + \int_D \frac{\partial \mathfrak{L}_\sigma(\eta,\zeta;\lambda)}{\partial \zeta_j} \mathfrak{W}(\eta) ds_\eta,$$

$$\frac{\partial \mathfrak{W}_\sigma(\zeta)}{\partial \zeta_j} = \int_\Sigma \frac{\partial L_\sigma(\eta,\zeta;\lambda)}{\partial \zeta_j} \mathfrak{W}(\eta) ds_\eta, \quad \zeta \in \Omega, \quad j = \overline{1,4}. \tag{5.22}$$

Taking into account (5.22) and (5.11), we obtain

$$\left| \frac{\partial \mathfrak{W}(\zeta)}{\partial \zeta_j} - \frac{\partial_\sigma \mathfrak{W}(\zeta)}{\partial \zeta_j} \right| \le \left| \int_D \frac{\partial \mathfrak{L}_\sigma(\eta,\zeta;\lambda)}{\partial \zeta_j} \mathfrak{W}(\eta) ds_\eta \right| \le \\ \le \int_D \left| \frac{\partial \mathfrak{L}_\sigma(\eta,\zeta;\lambda)}{\partial \zeta_j} \right| |\mathfrak{W}(\eta)| ds_\eta \le M \int_D \left| \frac{\partial \mathfrak{L}_\sigma(\eta,\zeta;\lambda)}{\partial \zeta_j} \right| ds_\eta, \\ \zeta \in \Omega, \quad j = \overline{1,4}. \tag{5.23}$$

Now let's estimate the integrals
$\int_D \left| \frac{\partial \Psi_\sigma(\eta,\zeta;\lambda)}{\partial \zeta_j} \right| ds_\eta$, $(j = \overline{1,3})$ and $\int_D \left| \frac{\partial \Psi_\sigma(\eta,\zeta;\lambda)}{\partial \zeta_m} \right| ds_\eta$ on the part D of the plane $\eta_4 = 0$.

To estimate the first integrals, we use the equality

$$\frac{\partial\Psi_\sigma(\eta,\zeta;\lambda)}{\partial\zeta_j}=\frac{\partial\Psi_\sigma(\eta,\zeta;\lambda)}{\partial s}\frac{\partial s}{\partial\zeta_j}=-2(\eta_j-\zeta_j)\frac{\partial\Psi_\sigma(\eta,\zeta;\lambda)}{\partial s}, \quad s=\alpha^2,\quad j=\overline{1,3}. \tag{5.24}$$

Given equality (5.16), inequality (5.17) and equality (5.24), we obtain

$$\int_D\left|\frac{\partial\Psi_\sigma(\eta,\zeta;\lambda)}{\partial\zeta_j}\right|ds_y\le\mathfrak{F}(\lambda,\zeta)\sigma^2e^{-\sigma\zeta_4},\quad\sigma>1,\quad\zeta\in\Omega, \quad j=\overline{1,3}. \tag{5.25}$$

Now we estimate the integral $\int_D\left|\frac{\partial\Psi_\sigma(\eta,\zeta;\lambda)}{\partial\zeta_4}\right|ds_\eta$.

Taking into account equality (5.16) and inequality (5.17), we obtain

$$\int_D\left|\frac{\partial\Psi_\sigma(\eta,\zeta;\lambda)}{\partial\zeta_4}\right|ds_\eta\le\mathfrak{F}(\lambda,\zeta)\sigma^2e^{-\sigma\zeta_4},\quad\zeta\in\Omega. \tag{5.26}$$

From inequalities (5.23), (5.25) and (5.26), we obtain the estimate (5.14). Theorem 3.1 is proved. □

Corollary 3.2. *For each $\zeta\in\Omega$, the equalities*

$$\lim_{\sigma\to\infty}\mathfrak{W}_\sigma(\zeta)=\mathfrak{W}(\zeta),\quad\lim_{\sigma\to\infty}\frac{\partial\mathfrak{W}_\sigma(\zeta)}{\partial\zeta_j}=\frac{\partial\mathfrak{W}(\zeta)}{\partial\zeta_j},\quad j=\overline{1,4},$$

are true.

We denote by $\overline{\Omega}_\varepsilon$ the set

$$\overline{\Omega}_\varepsilon=\left\{(\zeta_1,\ldots,\zeta_m)\in\zeta,\quad q>\zeta_m\ge\varepsilon,\quad q=\max_D\psi(\zeta'),\quad 0<\varepsilon<q\right\}.$$

Here $\psi(\zeta')$ is a surface. It is easy to see that the set $\overline{\Omega}_\varepsilon\subset\Omega$ is compact.

Corollary 3.3. *If $\zeta\in\overline{\Omega}_\varepsilon$, then the families of functions $\{\mathfrak{W}_\sigma(\zeta)\}$ and $\left\{\frac{\partial\mathfrak{W}_\sigma(\zeta)}{\partial\zeta_j}\right\}$ converge uniformly as $\sigma\to\infty$, i.e.:*

$$\mathfrak{W}_\sigma(\zeta)\rightrightarrows\mathfrak{W}(\zeta),\quad\frac{\partial\mathfrak{W}_\sigma(\zeta)}{\partial\zeta_j}\rightrightarrows\frac{\partial\mathfrak{W}(\zeta)}{\partial\zeta_j},\quad j=\overline{1,4}.$$

It should be noted that the set $E_\varepsilon=\Omega\backslash\overline{\Omega}_\varepsilon$ serves as a boundary layer for this problem, as in the theory of singular perturbations, where there is no uniform convergence.

4. Regularized solution of the problem (5.1)-(5.2)

Suppose that the surface Σ is given by the equation

$$\eta_m = \psi(\eta'), \quad \eta' \in \mathbb{R}^3,$$

where $\psi(\eta')$ is a single-valued function satisfying the Lyapunov conditions. We put

$$q = \max_D \psi(\eta'), \quad l = \max_D \sqrt{1 + \psi'^2(\eta')}.$$

Theorem 4.1. *Let $\mathfrak{W}(\eta) \in S(\Omega)$ satisfy condition (5.11), and the inequality*

$$|\mathfrak{W}(\eta)| \leq \delta, \quad 0 < \delta < 1, \tag{5.27}$$

hold on a smooth surface Σ. Then the following estimates are true:

$$|\mathfrak{W}(\zeta)| \leq \mathfrak{F}(\lambda,\zeta)\sigma^2 M^{1-\frac{\zeta_4}{q}} \delta^{\frac{\zeta_4}{q}}, \quad \sigma > 1, \quad \zeta \in \Omega. \tag{5.28}$$

$$\left|\frac{\partial \mathfrak{W}(\zeta)}{\partial \zeta_j}\right| \leq \mathfrak{F}(\lambda,\zeta)\sigma^2 M^{1-\frac{\zeta_4}{q}} \delta^{\frac{\zeta_4}{q}}, \quad \sigma > 1, \quad \zeta \in \Omega, \quad j = \overline{1,4}. \tag{5.29}$$

Proof. Let us first estimate inequality (5.28). Using the integral formula (5.10), we have

$$\mathfrak{W}(\zeta) = \int\limits_{\Sigma} \mathfrak{L}_\sigma(\eta,\zeta;\lambda)\mathfrak{W}(\eta)ds_\eta + \int\limits_{D} \mathfrak{L}_\sigma(\eta,\zeta;\lambda))\mathfrak{W}(\eta)ds_\eta, \quad \zeta \in \Omega. \tag{5.30}$$

We have

$$|\mathfrak{W}(\zeta)| \leq \left|\int\limits_{\Sigma} \mathfrak{L}_\sigma(\eta,\zeta;\lambda)\mathfrak{W}(\eta)ds_\eta\right| + \left|\int\limits_{D} \mathfrak{L}_\sigma(\eta,\zeta;\lambda)\mathfrak{W}(\eta)ds_\eta\right|, \quad \zeta \in \Omega. \tag{5.31}$$

Given inequality (5.27), let's estimate the first integral of inequality (5.31):

$$\left|\int\limits_{\Sigma} \mathfrak{L}_\sigma(\eta,\zeta;\lambda)W(\eta)ds_\eta\right| \leq \int\limits_{\Sigma} |\mathfrak{L}_\sigma(\eta,\zeta;\lambda)|\,|\mathfrak{W}(\eta)|ds_\eta \leq$$
$$\leq \delta \int\limits_{\Sigma} |L_\sigma(\eta,\zeta;\lambda)|ds_\eta, \quad \zeta \in \Omega. \tag{5.32}$$

Now let's estimate the integrals $\int\limits_{\Sigma} |\Psi_\sigma(\eta,\zeta;\lambda)|ds_\eta$, $\int\limits_{\Sigma} \left|\frac{\partial \Psi_\sigma(\eta,\zeta;\lambda)}{\partial \eta_j}\right| ds_\eta$, $(j=\overline{1,3})$ and $\int\limits_{\Sigma} \left|\frac{\partial \Psi_\sigma(\eta,\zeta;\lambda)}{\partial \eta_4}\right| ds_\eta$ on a smooth surface Σ.

Given equality (5.16) and inequality (5.17), we have

$$\int\limits_{\Sigma} |\Psi_\sigma(\eta,\zeta;\lambda)|ds_\eta \leq \mathfrak{F}(\lambda,\zeta)\sigma^2 e^{\sigma(q-\zeta_4)}, \quad \sigma>1, \quad \zeta\in\Omega. \tag{5.33}$$

To estimate the second integral, using equalities (5.16) and (5.19) as well as inequality (5.17), we obtain

$$\int\limits_{\Sigma} \left|\frac{\partial \Psi_\sigma(\eta,\zeta;\lambda)}{\partial \eta_j}\right| ds_\eta \leq \mathfrak{F}(\lambda,\zeta)\sigma^2 e^{\sigma(q-\zeta_4)}, \quad \sigma>1, \quad \zeta\in\Omega, \quad j=\overline{1,3}. \tag{5.34}$$

To estimate the integral $\int\limits_{\Sigma} \left|\frac{\partial \Psi_\sigma(\eta,\zeta;\lambda)}{\partial \eta_4}\right| ds_\eta$, using equality (5.16) and inequality (5.17), we obtain

$$\int\limits_{\Sigma} \left|\frac{\partial \Psi_\sigma(\eta,\zeta;\lambda)}{\partial \eta_4}\right| ds_\eta \leq \mathfrak{F}(\lambda,\zeta)\sigma^2 e^{\sigma(q-\zeta_4)}, \quad \sigma>1, \quad \zeta\in\Omega. \tag{5.35}$$

From (5.33)-(5.35), we obtain

$$\left|\int\limits_{\Sigma} \mathfrak{L}_\sigma(\eta,\zeta;\lambda)W(\eta)ds_\eta\right| \leq \mathfrak{F}(\lambda,\zeta)\sigma^2\delta e^{\sigma(q-\zeta_4)}, \quad \sigma>1, \quad \zeta\in\Omega. \tag{5.36}$$

The following is known:

$$\left|\int\limits_{D} \mathfrak{L}_\sigma(\eta,\zeta;\lambda)\mathfrak{W}(\eta)ds_\eta\right| \leq M\mathfrak{F}(\lambda,\zeta)\sigma^2 e^{-\sigma\zeta_4}, \quad \sigma>1, \quad \zeta\in\Omega. \tag{5.37}$$

Now taking into account (5.36)-(5.37), we have

$$|\mathfrak{W}(\zeta)| \leq \frac{\mathfrak{F}(\lambda,\zeta)\sigma^2}{2}(\delta e^{\sigma q}+M)e^{-\sigma\zeta_4}, \quad \sigma>1, \quad \zeta\in\Omega. \tag{5.38}$$

Choosing σ from the equality

$$\sigma = \frac{1}{q}\ln\frac{M}{\delta}, \tag{5.39}$$

we obtain the estimate (5.28).

Now let us prove inequality (5.29). To do this, we find the partial derivative from the integral formula (5.10) with respect to the variable ζ_j, $\quad j=\overline{1,3}$:

$$\frac{\partial \mathfrak{W}(\zeta)}{\partial \zeta_j} = \int\limits_{\Sigma} \frac{\partial \mathfrak{L}_\sigma(\eta,\zeta;\lambda)}{\partial \zeta_j}\mathfrak{W}(\eta)ds_\eta + \int\limits_{D} \frac{\partial \mathfrak{L}_\sigma(\eta,\zeta;\lambda)}{\partial \zeta_j}\mathfrak{W}(\eta)ds_\eta +$$

$$+\frac{\partial \mathfrak{W}_\sigma(\zeta)}{\partial \zeta_j} + \int\limits_{D} \frac{\partial \mathfrak{L}_\sigma(\eta,\zeta;\lambda)}{\partial \zeta_j}\mathfrak{W}(\eta)ds_\eta, \quad \zeta\in\Omega, \quad j=\overline{1,4}. \tag{5.40}$$

Here

$$\frac{\partial \mathfrak{W}_\sigma(\zeta)}{\partial \zeta_j} = \int\limits_{\Sigma} \frac{\partial \mathfrak{L}_\sigma(\eta,\zeta;\lambda)}{\partial \zeta_j}\mathfrak{W}(\eta)ds_\eta. \tag{5.41}$$

We have

$$\left|\frac{\partial \mathfrak{W}(\zeta)}{\partial \zeta_j}\right| \le \left|\int\limits_{\Sigma} \frac{\partial \mathfrak{L}_\sigma(\eta,\zeta;\lambda)}{\partial \zeta_j}\mathfrak{W}(\eta)ds_\eta\right| + \left|\int\limits_{D} \frac{\partial \mathfrak{L}_\sigma(\eta,\zeta;\lambda)}{\partial \zeta_j}\mathfrak{W}(\eta)ds_\eta\right| \le$$

$$\le \left|\frac{\partial \mathfrak{W}_\sigma(\zeta)}{\partial \zeta_j}\right| + \left|\int\limits_{D} \frac{\partial \mathfrak{L}_\sigma(\eta,\zeta;\lambda))}{\partial \zeta_j}\mathfrak{W}(\eta)ds_\eta\right|, \quad \zeta\in\Omega, \quad j=\overline{1,4}. \tag{5.42}$$

Given inequality (5.27), we estimate the first integral of inequality (5.42):

$$\left|\int\limits_{\Sigma} \frac{\partial \mathfrak{L}_\sigma(\eta,\zeta;\lambda)}{\partial \zeta_j}\mathfrak{W}(\eta)ds_\eta\right| \le \int\limits_{\Sigma} \left|\frac{\partial \mathfrak{L}_\sigma(\eta,\zeta;\lambda)}{\partial \zeta_j}\right| |\mathfrak{W}(\eta)|ds_\eta \le$$

$$\le \delta \int\limits_{\Sigma} \left|\frac{\partial \mathfrak{L}_\sigma(\eta,\zeta;\lambda)}{\partial \zeta_j}\right| ds_\eta, \quad \zeta\in\Omega, \quad j=\overline{1,4}. \tag{5.43}$$

To do this, we estimate the integrals $\int\limits_{\Sigma}\left|\frac{\partial \Psi_\sigma(\eta,\zeta;\lambda)}{\partial \zeta_j}\right| ds_\eta, \quad (j=\overline{1,3})$ and $\int\limits_{\Sigma}\left|\frac{\partial \Psi_\sigma(\eta,\zeta;\lambda)}{\partial \zeta_4}\right| ds_\eta$ on a smooth surface Σ.

Given equality (5.16), inequality (5.17) and equality (5.24), we obtain

$$\int\limits_{\Sigma} \left| \frac{\partial \Psi_{\sigma}(\eta,\zeta;\lambda)}{\partial \zeta_j} \right| ds_{\eta} \leq \mathfrak{F}(\lambda,\zeta)\sigma^2 e^{\sigma(q-\zeta_4)}, \quad \sigma > 1, \quad \zeta \in \Omega, \quad j=\overline{1,3}. \tag{5.44}$$

Now we estimate the integral $\int\limits_{\Sigma} \left| \frac{\partial \Psi_{\sigma}(\eta,\zeta;\lambda)}{\partial \zeta_4} \right| ds_{\eta}$.

Taking into account equality (5.16) and inequality (5.17), we obtain

$$\int\limits_{\Sigma} \left| \frac{\partial \Psi_{\sigma}(\eta,\zeta;\lambda)}{\partial \zeta_4} \right| ds_{\eta} \leq \mathfrak{F}(\lambda,\zeta)\sigma^2 e^{\sigma(q-\zeta_4)}, \quad \sigma > 1, \quad \zeta \in \Omega. \tag{5.45}$$

From (5.44)-(5.45), we obtain

$$\left| \int\limits_{\Sigma} \frac{\partial \mathfrak{L}_{\sigma}(\eta,\zeta;\lambda)}{\partial \zeta_j} \mathfrak{W}(\eta) \right| \leq K(\lambda,\zeta)\sigma^2 \delta e^{\sigma(q-\zeta_4)}, \quad \sigma > 1, \quad \zeta \in \Omega, \quad j=\overline{1,4}. \tag{5.46}$$

The following is known:

$$\left| \int\limits_{D} \frac{\partial \mathfrak{L}_{\sigma}(\eta,\zeta;\lambda)}{\partial \zeta_j} \mathfrak{W}(\eta) ds_{\eta} \right| \leq M\mathfrak{F}(\lambda,\zeta)\sigma^2 e^{-\sigma\zeta_4}, \quad \sigma > 1, \quad \zeta \in \Omega, \quad j=\overline{1,4}. \tag{5.47}$$

Now taking into account (5.46)-(5.47), bearing in mind (5.42), we have

$$\left| \frac{\partial \mathfrak{W}(\zeta)}{\partial \zeta_j} \right| \leq \frac{\mathfrak{F}(\lambda,\zeta)\sigma^2}{2}(\delta e^{\sigma q} + M)e^{-\sigma\zeta_4}, \quad \sigma > 1, \quad \zeta \in \Omega, \quad j=\overline{1,4}. \tag{5.48}$$

Choosing σ from the equality (5.39), we obtain the estimate (5.29).

Theorem 4.1 is proved. □

Let $\mathfrak{W}(\eta) \in S(\Omega)$ and let the approximation $f_{\delta}(\eta)$ be given instead of $\mathfrak{W}(\eta)$ on Σ with an error $0 < \delta < 1$,

$$\max_{\Sigma} |\mathfrak{W}(\eta) - f_{\delta}(\eta)| \leq \delta. \tag{5.49}$$

We put

$$\mathfrak{W}_{\sigma(\delta)}(\zeta) = \int_{\Sigma} \mathfrak{L}_\sigma(\eta,\zeta;\lambda) f_\delta(\eta) ds_y, \quad \zeta \in \Omega. \tag{5.50}$$

Theorem 4.2. *Let $W(\eta) \in S(\Omega)$ satisfy condition (5.11) on the part of the plane $\eta_m = 0$.*

Then the following estimates are true:

$$\left|\mathfrak{W}(\zeta) - \mathfrak{W}_{\sigma(\delta)}(\zeta)\right| \le \mathfrak{L}(\lambda,\zeta)\sigma^2 M^{1-\frac{\zeta_4}{q}} \delta^{\frac{\zeta_4}{q}}, \quad \sigma > 1, \quad \zeta \in \Omega. \tag{5.51}$$

$$\left|\frac{\partial \mathfrak{W}(\zeta)}{\partial \zeta_j} - \frac{\partial \mathfrak{W}_{\sigma(\delta)}(\zeta)}{\partial \zeta_j}\right| \le K(\lambda,\zeta)\sigma^2 M^{1-\frac{\zeta_4}{q}} \delta^{\frac{\zeta_4}{q}}, \quad \sigma > 1, \quad \zeta \in \Omega, \quad j = \overline{1,4}. \tag{5.52}$$

Proof. From the integral formulas (5.10) and (5.50), we have

$$\mathfrak{W}(\zeta) - \mathfrak{W}_{\sigma(\delta)}(\zeta) = \int_{\partial\Omega} L_\sigma(\eta,\zeta;\lambda)\mathfrak{W}(\eta) ds_\eta - \int_{\Sigma} \mathfrak{L}_\sigma(\eta,\zeta;\lambda) f_\delta(\eta) ds_\eta =$$

$$= \int_{\Sigma} L_\sigma(\eta,\zeta;\lambda)\mathfrak{W}(\eta) ds_\eta + \int_{D} \mathfrak{L}_\sigma(\eta,\zeta;\lambda)\mathfrak{W}(\eta) ds_\eta - \int_{\Sigma} \mathfrak{L}_\sigma(\eta,\zeta;\lambda) f_\delta(\eta) ds_\eta =$$

$$= \int_{\Sigma} \mathfrak{L}_\sigma(\eta,\zeta;\lambda)\left\{\mathfrak{W}(\eta) - f_\delta(\eta)\right\} ds_\eta + \int_{D} L_\sigma(\eta,\zeta;\lambda)\mathfrak{W}(\eta) ds_\eta.$$

and

$$\frac{\partial \mathfrak{W}(\zeta)}{\partial \zeta_j} - \frac{\partial \mathfrak{W}_{\sigma(\delta)}(\zeta)}{\partial \zeta_j} = \int_{\partial\Omega} \frac{\partial \mathfrak{L}_\sigma(\eta,\zeta;\lambda)}{\partial \zeta_j}\mathfrak{W}(\eta) ds_\eta - \int_{\Sigma} \frac{\partial \mathfrak{L}_\sigma(\eta,\zeta;\lambda)}{\partial \zeta_j} f_\delta(\eta) ds_\eta =$$

$$= \int_{\Sigma} \frac{\partial \mathfrak{L}_\sigma(\eta,\zeta;\lambda)}{\partial \zeta_j}\mathfrak{W}(\eta) ds_\eta + \int_{D} \frac{\partial \mathfrak{L}_\sigma(\eta,\zeta;\lambda)}{\partial \zeta_j}\mathfrak{W}(\eta) ds_\eta - \int_{\Sigma} \frac{\partial \mathfrak{L}_\sigma(\eta,\zeta;\lambda)}{\partial \zeta_j} f_\delta(\eta) ds_\eta =$$

$$= \int_{\Sigma} \frac{\partial \mathfrak{L}_\sigma(\eta,\zeta;\lambda)}{\partial \zeta_j}\left\{\mathfrak{W}(\eta) - f_\delta(\eta)\right\} ds_\eta + \int_{D} \frac{\partial \mathfrak{L}_\sigma(\eta,\zeta;\lambda)}{\partial \zeta_j}\mathfrak{W}(\eta) ds_\eta, \quad j = \overline{1,4}.$$

Using conditions (5.11) and (5.49), we obtain

$$\left|\mathfrak{W}(\zeta) - \mathfrak{W}_{\sigma(\delta)}(\zeta)\right| = \left|\int_{\Sigma} \mathfrak{L}_\sigma(\eta,\zeta;\lambda)\left\{\mathfrak{W}(\eta) - f_\delta(\eta)\right\} ds_\eta\right| +$$

$$+ \left|\int_{D} \mathfrak{L}_\sigma(\eta,\zeta;\lambda)\mathfrak{W}(\eta) ds_\eta\right| \le \int_{\Sigma} |\mathfrak{L}_\sigma(\eta,\zeta;\lambda)|\,|\{\mathfrak{W}(\eta) - f_\delta(\eta)\}|\, ds_\eta +$$

$$+ \int_{D} |\mathfrak{L}_\sigma(\eta,\zeta;\lambda)|\,|\mathfrak{W}(\eta)|\, ds_\eta \le \delta \int_{\Sigma} |\mathfrak{L}_\sigma(\eta,\zeta;\lambda)|\, ds_\eta + M \int_{D} |\mathfrak{L}_\sigma(\eta,\zeta;\lambda)|\, ds_\eta.$$

and

$$\left|\frac{\partial\mathfrak{W}(\zeta)}{\partial\zeta_j}-\frac{\partial\mathfrak{W}_{\sigma(\delta)}(\zeta)}{\partial\zeta_j}\right|=\left|\int\limits_{\Sigma}\frac{\partial\mathfrak{L}_{\sigma}(\eta,\zeta;\lambda)}{\partial\zeta_j}\{\mathfrak{W}(\eta)-f_{\delta}(\eta)\}\,ds_y\right|+$$

$$+\left|\int\limits_{D}\frac{\partial\mathfrak{L}_{\sigma}(\eta,\zeta;\lambda)}{\partial\zeta_j}\mathfrak{W}(\eta)ds_{\eta}\right|\leq\int\limits_{\Sigma}\left|\frac{\partial\mathfrak{L}_{\sigma}(\eta,\zeta;\lambda)}{\partial\zeta_j}\right||\{\mathfrak{W}(\eta)-f_{\delta}(\eta)\}|ds_{\eta}+$$

$$+\int\limits_{D}\left|\frac{\partial\mathfrak{L}_{\sigma}(\eta,\zeta;\lambda)}{\partial\zeta_j}\right||\mathfrak{W}(\eta)|ds_{\eta}\leq\delta\int\limits_{\Sigma}\left|\frac{\partial\mathfrak{L}_{\sigma}(\eta,\zeta;\lambda)}{\partial\zeta_j}\right|ds_{\eta}+$$

$$+M\int\limits_{D}\left|\frac{\partial\mathfrak{L}_{\sigma}(\eta,\zeta;\lambda)}{\partial\zeta_j}\right|ds_{\eta},\quad j=\overline{1,4}.$$

Now, repeating the proofs of Theorems 3.1 and 4.1, we obtain

$$\left|\mathfrak{W}(\zeta)-\mathfrak{W}_{\sigma(\delta)}(\zeta)\right|\leq\frac{\mathfrak{F}(\lambda,\zeta)\sigma^2}{2}(\delta e^{\sigma q}+M)e^{-\sigma\zeta_4}.$$

$$\left|\frac{\partial\mathfrak{W}(\zeta)}{\partial\zeta_j}-\frac{\mathfrak{W}_{\sigma(\delta)}(\zeta)}{\partial\zeta_j}\right|\leq\frac{\mathfrak{F}(\lambda,\zeta)\sigma^2}{2}(\delta e^{\sigma q}+M)e^{-\sigma\zeta_4},\quad j=\overline{1,4}.$$

From here, choosing σ from equality (5.39), we have the estimates (5.51) and (5.52).

Theorem 4.2 is proved. □

Corollary 4.3. *For each $\zeta\in\Omega$, the following equalities are true:*

$$\lim_{\delta\to 0}\mathfrak{W}_{\sigma(\delta)}(\zeta)=\mathfrak{W}(\zeta),\quad \lim_{\delta\to 0}\frac{\partial\mathfrak{W}_{\sigma(\delta)}(\zeta)}{\partial\zeta_j}=\frac{\partial\mathfrak{W}(\zeta)}{\partial\zeta_j},\quad j=\overline{1,4}.$$

Corollary 4.4. *If $x\in\overline{\Omega}_{\varepsilon}$, then the families of functions $\{\mathfrak{W}_{\sigma(\delta)}(\zeta)\}$ and $\left\{\frac{\partial\mathfrak{W}_{\sigma(\delta)}(\zeta)}{\partial\zeta_j}\right\}$ converge uniformly as $\delta\to 0$, i.e.:*

$$\mathfrak{W}_{\sigma(\delta)}(\zeta)\rightrightarrows\mathfrak{W}(\zeta),\quad \frac{\partial\mathfrak{W}_{\sigma(\delta)}(\zeta)}{\partial\zeta_j}\rightrightarrows\frac{\partial\mathfrak{W}(\zeta)}{\partial\zeta_j},\quad j=\overline{1,4}.$$

5. Conclusion

In this chapter, we have explicitly found a regularized solution to the ill-posed Cauchy problem for matrix factorizations of the Helmholtz equation in a four-dimensional bounded domain. It is assumed that a solution to the problem exists and is continuously differentiable in a closed domain with exactly given Cauchy data. For this case, an explicit formula for the continuation of the solution is established, as well as a regularization formula for the situation where, under the indicated conditions, instead of the Cauchy data, their continuous approximations with a given error in the uniform metric are provided. We have obtained a stability estimate for the solution of the Cauchy problem in the classical sense.

An estimate for the stability of the solution of the Cauchy problem in the classical sense for matrix factorizations of the Helmholtz equation is given. A problem is considered where instead of exact data of the Cauchy problem, their approximations with a given deviation in the uniform metric are given and under the assumption that the solution of the Cauchy problem is bounded on part D of the boundary of the domain Ω, an explicit regularization formula is obtained.

We note that when solving applied problems, one should find the approximate values of $\mathfrak{W}(\zeta)$ and $\dfrac{\partial \mathfrak{W}(\zeta)}{\partial \zeta_j}$, $\zeta \in \Omega$, $j = \overline{1,4}$.

Also, we construct a family of vector-functions $\mathfrak{W}(\zeta, f_\delta) = \mathfrak{W}_{\sigma(\delta)}(\zeta)$ and $\dfrac{\partial \mathfrak{W}(\zeta, f_\delta)}{\partial \zeta_j} = \dfrac{\partial \mathfrak{W}_{\sigma(\delta)}(\zeta)}{\partial \zeta_j}$, $j = \overline{1,4}$ depending on a parameter σ, and prove that under certain conditions and a special choice of the parameter $\sigma = \sigma(\delta)$, as $\delta \to 0$, the family $\mathfrak{W}_{\sigma(\delta)}(\zeta)$ and $\dfrac{\partial \mathfrak{W}_{\sigma(\delta)}(\zeta)}{\partial \zeta_j}$ converges in the usual sense to a solution $\mathfrak{W}(\zeta)$ and its derivative $\dfrac{\partial \mathfrak{W}(\zeta)}{\partial \zeta_j}$ at the point $\zeta \in \Omega$. Following A.N. Tikhonov (see [197]), the family of vector-valued functions $\mathfrak{W}_{\sigma(\delta)}(\zeta)$ and $\dfrac{\partial \mathfrak{W}_{\sigma(\delta)}(\zeta)}{\partial \zeta_j}$ is called a regularized solution of the problem. A regularized solution provides a stable method for approximately solving the problem.

Thus, functionals $\mathfrak{W}_{\sigma(\delta)}(\zeta)$ and $\dfrac{\partial \mathfrak{W}_{\sigma(\delta)}(\zeta)}{\partial \zeta_j}$ determine the regularization of the solution of problem (5.1)-(5.2).

Chapter 6

Conflict-controlled Systems in Different Spaces

B.T. Samatov[1] *
M.A. Turgunboeva[2]
[1]V.I. Romanovsky Institute of Mathematics at the Academy of Sciences of the Republic of Uzbekistan, Tashkent, Uzbekistan
[2]Namangan State University, Namangan, Uzbekistan

Abstract

This chapter examines two conflicting controlled objects, the movements of which are inertial in nature. The goal of the first object (the pursuer) is to l-capture the second object (the evader), that is, to approach the enemy at a distance of l. The goal of the second object is the opposite, that is, to prevent a pursuing approach to it at a distance of l. The work consists of two parts. In the first part, the control of each participant is selected from the space L_∞, that is, from the class of measurable bounded functions. In the second part, the control of each participant is selected from the L_2 class, that is, from the class of square summable functions. For each case, when solving pursuit problems, a strategy for the best approach is constructed. Necessary and sufficient conditions for each case when solving pursuit problems are obtained. The work has theoretical and applied significance in the field of control theory and is a continuation of the works of Isaacs, Azimov and Guseynov, Pshenichnii, Satimov, Azamov, the authors of this article, and others.

*Corresponding Author's Email: samatov57@gmail.com

Keywords: differential game, l-capture, acceleration, strategy, guaranteed time

AMS Subject Classification: 49N70, 49N75, 91A23, 91A24.

1. Introduction

Control problem $\dot{z} = F(z,u)$, and a more general theory of controlled conflict processes described by an equation of the form $\dot{z} = F(z,u,v)$ is usually considered under a geometric constraint imposed on the control vectors of the form $u \in \mathbf{P}, v \in \mathbf{Q}$, where $\mathbf{P}$ and $\mathbf{Q}$ are the given subsets of Euclidean spaces of the corresponding dimension, i.e. the managing subject must choose a control method that generates a measurable function $u(t), 0 \le t \le t^*$ (respectively $v(t), 0 \le t \le t^*$), such that

$$u(t) \in \mathbf{P} \text{ a.e. } \ (v(t) \in \mathbf{Q} \text{ a.e.})$$

For example, if $\mathbf{P}$ is a ball $|u| \le \varsigma$, then geometric constraint is equivalent to the requirement

$$\|u(\cdot)\|_\infty = \operatorname*{ess\ sup}_{0 \le t \le t^*} |u(t)| \le \varsigma.$$

From a functional point of view, a natural analogue of this condition is the integral constraint

$$\|u(\cdot)\|_p = \left[\int_0^{t^*} |u(t)|^p dt \right]^{1/p} \le \varsigma, \ \ \varsigma > 0, \ \ p \ge 1.$$

Problems with integral constraints are more complex than problems with geometric constraints, but in practice, both types of constraints are important: while the first type expresses the limited dynamic capabilities of the object (for example, a restriction on traction force), the second type expresses the finiteness of resources (for example, fuel).

Problems of conflicting controllable objects described by ordinary differential equations began to be systematically studied by the American mathematician Isaacs in the 1950s. Isaacs's research [264] was published in the form of a monograph, which included a large number of problems in the theory of conflict-controlled processes. He called this theory the theory of differential games.In solving differential game problems, the author primarily relied on the classical theory of calculus of variations and made significant efforts to apply the Hamilton-Jacobi-Belman method (known as the Isaacs method).

Foundation of the theory of differential games was settled by Fleming [259], Friedman [260], Pshenichnii [272], Subbotin [273], Subbotin and Chentsov [274], Berkovitz [257], Petrosyan [269], Pontryagin [271], and others. Based on the fundamental approaches advanced by Krasovsky and Subbotin [266], and Pontryagin [271] a differential game is viewed as a control problem from the standpoint of either a pursuer or an evader. In accordance with this consideration, the game is reduced to either pursuit (approach) problem or to evasion (escape) problem. Analogous effects of the pursuit-evasion problems in the cases of geometric and integral constraints were considered by other scientists (Azamov [253]; Azimov [255]; Azimov and Guseynov [256]; Petrov [270]; Vagin and Petrov [282]; Ibragimov and Salimi [263]; Azamov and Samatov [254]; Samatov [275]– [277]).

The problem for the case of l-approach (Nahin [267]) was first studied by Indian mathematician Ramchundra. He considered an intercept problem involving a slow pursuer and a fast target. In the work of Petrosyan and Dutkevich [268], the l-capture problem was considered for the players moving at the limited velocities by the coordinates on the plane and also, a "Life-line" game (Problem 9.5.1, Isaacs [264]) was solved by geometrical method.

In the work of Khaidarov [265], considered the problem of positional l-capture of one evader by a group of pursuers provided that each of the players has a simple movement. Grigorenko [261] solved the l-capture problem for Pontryagin's control example by exchanging given differential equations for a linear normal system. Satimov [281] investigated the l-capture problem when players were moving in a convex compact set, along a convex surface, and at the edges of the cube (the length of the edge is n). Samatov [277] solved the problem of group pursuit for the case of l-catch in simple motion of the players, with integral constraints imposed on controls, on the basis of Chikrii's method of resolving functions. Samatov and et.al. [280] studied Ramchundra's problem for the l-capture problem at a lower pursuer speed. In this paper [280], a strategy was proposed that allows a boat to approach a ship moving at a faster speed than the boat. l_∞-escape problems in a broad class of linear differential games with many pursuers and one evader under integral constraints on controls were examined and sufficient conditions for these problems were obtained (Gusyatnikov and Mokhon'ko, [262]).

In the present chapter, we discuss the l-capture problems in a differential game with inertial players, whose controls are subject to both geometrical and integral constraints, and who have identical initial velocity vectors in Euclidean space. The work consists of two parts. In the first part, the control of each participant is selected from the space L_∞, that is, from the class of measurable bounded functions. In the second part, the control of each participant is selected from the L_2 class, that is, from the class of square summable functions. For each case, when solving pursuit problems, a strategy for the best approach is constructed and necessary and sufficient conditions are obtained. In addition, the shortest guaranteed l-capture time is presented.

2. Statement of Problems

In the Euclidean space $\mathbf{R}^n$, $n \geq 1$, two controlled objects move – the Pursuer P and the Evader E. Let x and y denote the location of objects P and E in this space, respectively. The motions of these objects are subject to the following differential equations:

$$P: \ \ddot{x} = u, \ \ x(0) = x_0, \ \ \dot{x}(0) = x_1, \tag{6.1}$$

$$E: \ \ddot{y} = v, \ \ y(0) = y_0, \ \ \dot{y}(0) = y_1, \tag{6.2}$$

respectively, where $x, y, u, v \in \mathbf{R}^n$; x_0, y_0 are the initial positions of the objects which are regarded as $|x_0 - y_0| > l$ for some $l > 0$ at the time $t = 0$; x_1, y_1 are their initial velocity vectors, respectively; the control parameters u, v are considered as objects' acceleration vectors depending on the time $t \geq 0$ and they are taken as measurable functions $u(\cdot) : [0, +\infty) \to \mathbf{R}^n$ and $v(\cdot) : [0, +\infty) \to \mathbf{R}^n$, respectively.

First of all, it is claimed that the control parameters u, v are subjected to the constraints

$$\operatorname*{ess\ sup}_{0 \leq t \leq t^*} |u(t)| \leq \alpha, \tag{6.3}$$

$$\operatorname*{ess\ sup}_{0 \leq t \leq t^*} |v(t)| \leq \beta, \tag{6.4}$$

which are usually termed geometric constraints (in short, G-constraint), where α, β are the given positive parametric numbers which represent the maximal acceleration values of the objects and t^* is the approach time of the objects at the distance l. It is worth noting that as the norms of $u(\cdot)$ and $v(\cdot)$ in (6.3) and (6.4) we refer to the Euclidean norms in the form

$$|u(t)| = \sqrt{\sum_{i=1}^{n} u_i^2(t)}, \quad |v(t)| = \sqrt{\sum_{i=1}^{n} v_i^2(t)}.$$

The family of all the measurable functions satisfying (6.3) is denoted by $\mathscr{U}_G$. Similarly, the family of all the measurable functions satisfying (6.4) is denoted by $\mathscr{V}_G$.

Likewise, we will later discuss the differential game in which the control parameters u and v in equations (6.1), (6.2) obey the the following type of constraints:

$$\int_0^t (t-s)|u(s)|^2 ds \leq \rho^2, \tag{6.5}$$

$$\int_0^t (t-s)|v(s)|^2 ds \leq \sigma^2, \tag{6.6}$$

where ρ and σ are the positive parametric numbers denoting the maximal resource values of the objects. In the theory of differential games, the inequalities (6.5) and (6.6) are usually called the integral constraints (in short, I-constraint). Usually, such constraints are termed L_2 classes, that is, from the classes of square summable functions.

We assign a class of all the measurable functions which satisfy the constraint (6.5) to $\mathscr{U}_I$. In the same way, we assign a class of all the measurable functions which satisfy the constraint (6.6) to $\mathscr{V}_I$.

Definition 2.1. *We say that quantities*

$$\rho(t) = \rho_0 - \int_0^t |u(s)|^2 ds, \quad \rho(0) = \rho_0^2,$$

$$\sigma(t) = \sigma_0 - \int_0^t |v(s)|^2 ds, \quad \sigma(0) = \sigma_0^2$$

are the residual resource of the pursuer and the evader at the current time t, $t \geqslant 0$.

Studying the problems associated with inertial players is more challenging than analyzing a simple pursuit differential game. These problems have emerged as a crucial area of research due to their wide-ranging applications in fields such as economics, engineering, and environmental science. There are many applications of differential games with inertial players under various restrictions on controls (Petrov [270]; Vagin and Petrov [282]; Samatov and Uralova [279]). When $l = 0$, a comprehensive analysis of the pursuit problem is studied for the differential game (6.1)–(6.4) and it is shown that in the case of inertial motions of players, the set of meeting points of the players is a linear combination of two such Apollonius balls, the first of which is built from the initial states, and the second from the initial states of the velocities (Samatov and Soyibboev [278]).

Definition 2.2. *The measurable function $u(\cdot) \in \mathscr{U}_G$ (or $u(\cdot) \in \mathscr{U}_I$) is called an admissible control of the object P.*

Definition 2.3. *The measurable function $v(\cdot) \in \mathscr{V}_G$ (or $v(\cdot) \in \mathscr{V}_I$) is called an admissible control of the object E.*

It should be noted that, the pair $(\mathscr{U}_G, \mathscr{V}_G)$ of the introduced families defines the G_l-Game. Just like, the pair $(\mathscr{U}_I, \mathscr{V}_I)$ of the introduced classes characterizes the I_l-Game. In the next steps, we will outline the general statements for both games.

If $u(\cdot) \in \mathscr{U}_G \left(u(\cdot) \in \mathscr{U}_I\right)$ and $v(\cdot) \in \mathscr{V}_G \left(v(\cdot) \in \mathscr{V}_I\right)$, then references to equations (6.1) and (6.2) show that each of the triplets $(x_0, x_1, u(\cdot))$ and $(y_0, y_1, v(\cdot))$ generates the motion trajectories

$$x(t) = x_0 + x_1 t + \int_0^t (t-s)u(s)ds,$$
$$y(t) = y_0 + y_1 t + \int_0^t (t-s)v(s)ds,$$

of the objects, respectively.

The basic goal of the pursuer is to equalize the distance with the evader to l $\big($the l-capture problem [258]$\big)$, viz, to achieve the relation

$$|x(t^*) - y(t^*)| \leq l \tag{6.7}$$

at some finite time $t^* > 0$. Whereas the objective of the evader is to avoid occurrence of (6.7), i.e., to keep the inequality $|x(t) - y(t)| > l$ for all $t \geq 0$ or, if it is impossible, to put back occurrence of (6.7).

Differential games G_l and I_l, with the help of program strategies which are only associated with the time $t \geq 0$, will not be solved to the advantage of the pursuer and the acceptable types of controls should be strategies. Therefore, we will present a more general definition for a strategy of the pursuer.

First of all, to be more comfortable subsequently, let's introduce the following denotations:

$$z = x - y, \; z_0 = x_0 - y_0, \; z_1 = x_1 - y_1.$$

Then the equations (6.1), (6.2) reduce to the Cauchy problem in the form

$$\ddot{z} = u - v, \; z(0) = z_0, \; \dot{z}(0) = z_1. \tag{6.8}$$

where z_0 is the difference between the initial positions, and z_1 is the difference between the initial velocities of the objects. In what follows, the vector z will be called the current state of the game, z_0 will be called the initial state of the game, and z_1 will be called the initial state of the velocity of the game. Because of that, when choosing admissible controls $u(\cdot) \in \mathscr{U}_G \left(u(\cdot) \in \mathscr{U}_I\right)$ and $v(\cdot) \in \mathscr{V}_G \left(v(\cdot) \in \mathscr{V}_I\right)$, the solution of equation (6.8) has the form

$$z(t) = z_0 + z_1 t + \int_0^t (t-s)(u(s) - v(s))ds. \tag{6.9}$$

Now, with the new denotations introduced above, the objective of the pursuer is to achieve the relation $|z(t)| \leq l$ from the given initial states z_0 and z_1 in the shortest possible time t, and the goal of the evader is to keep the inequality $|z(t)| > l$ for all $t \geq 0$.

Definition 2.4. *A map* $\mathbf{u}_G^* : \mathbf{R}^n \setminus lS \times \mathscr{V}_G \to \mathscr{U}_G \left(\mathbf{u}_I^* : \mathbf{R}^n \setminus lS \times \mathscr{V}_I \to \mathscr{U}_I\right)$ *is said to be a strategy of the pursuer if the following conditions are fulfilled:*

i) An inclusion $\mathbf{u}_G^*(z_0, v(\cdot)) \in \mathscr{U}_G \left(\mathbf{u}_I^*(z_0, v(\cdot)) \in \mathscr{U}_I\right)$ *is satisfied for each control* $v(\cdot) \in \mathscr{V}_G \left(v(\cdot) \in \mathscr{V}_I\right)$ *on some finite time interval* $[0,t]$*;*

ii) If for each $v_1(\cdot), v_2(\cdot) \in \mathscr{V}_G \left(v_1(\cdot), v_2(\cdot) \in \mathscr{V}_I\right)$ *and* $t \geq 0$*, the equality* $v_1(s) = v_2(s)$ *holds for almost everywhere on* $[0,t]$*, then* $u_1(s) = u_2(s)$ *is valid for almost everywhere on* $[0,t]$*, where* $u_i(\cdot) = \mathbf{u}_G^*(z_0, v_i(\cdot)) \left(u_i(\cdot) = \mathbf{u}_I^*(z_0, v_i(\cdot))\right)$, $i = 1, 2$ *, and* $S = \{s \in \mathbf{R}^n : |s| \leq 1\}$.

The present work is focused on discussing the l-capture problems with objects' motion dynamics (6.1)-(6.2) under the constraints (6.3)-(6.4), namely the G_l-Games and under the constraints the constraints (6.5)-(6.6), namely the I_l-Games. We define pursuer's strategy guaranteeing to be in the l-neighbourhood of the evader, and obtain necessary and sufficient conditions for each case when solving pursuit problems.

3. The G_l-Game

In many mathematical problems involving parameters, a notable characteristic of the eventual analytical outcomes is their clear reliance on these parameters, even though they are initially treated as constants in the solution. Nevertheless, these parameters play a crucial role in establishing the feasibility conditions for these problems. In this section, we will delineate the essential and comprehensive feasibility conditions for the l-capture problem within the game $\left(\mathscr{U}_G, \mathscr{V}_G\right)$. We defined this game as the G_l-Game where objects' motion dynamics are given by differential equations (6.1)-(6.2) under the geometric constraints in the form of (6.3)-(6.4).

For object P, only program strategies, which are admissible controls depending only on time t, are not enough to achieve its objective. To tackle the ℓ-capture problem, assume that at the current time t the pursuer possesses knowledge of the initial parameters z_0, z_1 as well as the constants α, β and l. Furthermore, the pursuer is informed about the current time t and is aware of the evader's control value $v(t)$. Thus, we can define the strategy of pursuer P in the form of dependence only on the current positions of the acceleration function $v(t)$, $t \geq 0$ and the given constants l, z_0, α.

Definition 3.1. *For $\alpha \geq \beta$, we call the function*

$$\mathbf{u}^*_{G_l}(z_0, v) = v - \lambda(z_0, v)\frac{\alpha z_0 + vl}{\alpha + \lambda(z_0, v)l} \tag{6.10}$$

the approach strategy or the Π_{G_l}-strategy of pursuer P in the G_l-Game (6.1)-(6.4), where

$$\lambda(z_0, v) = \mu + \sqrt{\mu^2 + \frac{\alpha^2 - |v|^2}{|z_0|^2 - l^2}}, \quad \mu = \frac{\langle v, z_0\rangle + \alpha l}{|z_0|^2 - l^2}. \tag{6.11}$$

$\langle v, z_0\rangle$ is the scalar product of the vectors v and z_0 in $\mathbf{R}^n$ and, moreover, the function $\lambda(z_0, v)$ is usually termed *the resolving function.*

Below we will provide some important properties for the strategy (6.10) and the resolving function (6.11).

Proposition 3.1. *If $\alpha \geq \beta$, then (6.10) is defined and continuous for any $v(\cdot)$, $v(\cdot) \in \mathscr{V}_G$, and the equality $|\mathbf{u}^*_{G_l}(z_0, v(\cdot))| = \alpha$ holds during the l-capture problem.*

Proposition 3.2. *If $\alpha \geq \beta$, then (6.11) is defined, non-negative and continuous for any $v(\cdot)$, $v(\cdot) \in \mathscr{V}_G$, and it is bounded as*

$$\frac{\alpha - \beta}{|z_0| - l} \leq \lambda(z_0, v(\cdot)) \leq \frac{\alpha + \beta}{|z_0| - l}.$$

We analyze the game for a more general case, where the vector z_1 is parallel to the vector z_0. In other words, there exists a real number k, $k \in \mathbf{R}$, such that $z_1 = kz_0$.

If the pursuer applies the Π_{G_l}-strategy (6.10) while the evader utilizes an optional admissible control $v(\cdot) \in \mathscr{V}_G$, then, according to the equality $z_1 = kz_0$ and depending on (6.9) we obtain the solution

$$z(t) = z_0 + z_0 kt + \int_0^t (t-s)(\mathbf{u}^*_{G_l}(z_0, v(s)) - v(s))ds. \tag{6.12}$$

Consider the function

$$\Lambda(t, v(\cdot)) = 1 + kt - \alpha\int_0^t (t-s)\frac{\lambda(z_0, v(s))}{\alpha + \lambda(z_0, v(s))l}ds \tag{6.13}$$

with respect to t.

Lemma 3.3. *Let $\alpha > \beta$ be valid. Then for any control $v(\cdot) \in \mathscr{V}_G$,*

a) if $k < 0$, then (6.13) is monotone decreasing with respect to t, $t \in [0,\infty]$ and, if $k \geq 0$, then (6.13) is monotone increasing up to the time τ and is monotone decreasing with respect to t, $t \in (\tau,\infty]$. Here, τ is the root of the equation

$$\int_0^t \frac{\lambda(z_0, v(s))ds}{\alpha + \lambda(z_0, v(s))l} = \frac{k}{\alpha};$$

b) (6.13) is bounded on time interval $[0,\theta]$ as follows:

$$\Lambda_1(t) \leq \Lambda(t, v(\cdot)) \leq \Lambda_2(t),$$

where

$$\Lambda_1(t) = 1 + kt - \frac{\alpha(\alpha+\beta)t^2}{2(\alpha|z_0| + \beta l)}, \quad \Lambda_2(t) = 1 + kt - \frac{\alpha(\alpha-\beta)t^2}{2(\alpha|z_0| - \beta l)}$$

and the number θ, which equals

$$\theta = \frac{k(\alpha|z_0| - \beta l) + \sqrt{k^2(\alpha|z_0| - \beta l)^2 + 2\alpha(\alpha-\beta)(\alpha|z_0| - \beta l)}}{\alpha(\alpha-\beta)}, \tag{6.14}$$

is the positive root of the equation $\Lambda_2(t) = 0$.

Proof. *a*) Let's compute the t-derivative of $\Lambda(t, v(\cdot))$ and we form

$$\frac{d\Lambda(t, v(\cdot))}{dt} = k - \alpha \int_0^t \frac{\lambda(z_0, v(s))ds}{\alpha + \lambda(z_0, v(s))l}. \tag{6.15}$$

It is clear that $d\Lambda(0, v(\cdot))/dt = k$, and also the right-hand side of (6.15) takes a negative value if $k < 0$. If we calculate the t-derivative of both sides of (6.15) again, then we obtain

$$\frac{d^2\Lambda(t, v(\cdot))}{dt^2} = -\frac{\alpha\lambda(z_0, v(t))}{\alpha + \lambda(z_0, v(t))L} \leq 0. \tag{6.16}$$

Depending on (6.16), $d\Lambda(t, v(\cdot))/dt \geq 0$, i.e., the function $\Lambda(t, v(\cdot))$ increases monotonically on some interval $[0, \tau]$, and $d\Lambda(t, v(\cdot))/dt < 0$, i.e., the function $\Lambda(t, v(\cdot))$ decreases monotonically on the interval $(\tau, +\infty)$ for all $k \geq 0$.

b) Let

$$f(w) = \frac{w}{\alpha + wl}, \tag{6.17}$$

where $\dfrac{\alpha-\beta}{|z_0|-l} \le w \le \dfrac{\alpha+\beta}{|z_0|-l}$.

According to monotonicity of the function (6.17) we have

$$\min f(w) = \frac{\alpha-\beta}{\alpha|z_0|-\beta l}, \tag{6.18}$$

$$\max f(w) = \frac{\alpha+\beta}{\alpha|z_0|+\beta l}. \tag{6.19}$$

By the lemma on minimum in the optimal control theory (pp.360, [252]) and from (6.18), we can write the following estimations down:

$$\Lambda(t,v(\cdot)) \le 1+kt - \min_{v(\cdot)\in\mathscr{V}_G} \int_0^t (t-s)\frac{\alpha\lambda(z_0,v(s))}{\alpha+\lambda(z_0,v(s))L}ds =$$

$$= 1+kt-\frac{\alpha t^2}{2}\min_{|v|\le\beta} f(\lambda(z_0,v)) \le \Lambda_2(t).$$

Similarly, by virtue of the identical lemma mentioned above and from (6.19), we obtain the following:

$$\Lambda_1(t) = 1+kt-\frac{\alpha t^2}{2}\max_{|v|\le\beta} f(\lambda(z_0,v)) =$$

$$= 1+kt - \max_{v(\cdot)\in\mathscr{V}_G} \int_0^t (t-s)\frac{\alpha\lambda(z_0,v(s))}{\alpha+\lambda(z_0,v(s))L}ds \le \Lambda(t,v(\cdot)).$$

The proof is finished. □

Take and examine the function

$$F(t) = -\frac{\alpha-\beta}{2}t^2 + |z_0|kt + |z_0| - l \tag{6.20}$$

under the conditions $\alpha=\beta$, $k<0$ or $\alpha>\beta$, $k\in\mathbf{R}$. Obviously, the equation $F(t)=0$ has a positive root with respect to t, $t\ge 0$ for these occasions. We will denote its positive root as T_l, where

$$T_l = \begin{cases} \left(|z_0|k+\sqrt{|z_0|^2k^2+2(\alpha-\beta)(|z_0|-l)}\right)/(\alpha-\beta) & \text{if } k\ne 0,\ \alpha>\beta, \\ (l-|z_0|)/|z_0|k & \text{if } k<0,\ \alpha=\beta, \\ \sqrt{2(|z_0|-l)/(\alpha-\beta)} & \text{if } k=0,\ \alpha>\beta. \end{cases} \tag{6.21}$$

Lemma 3.4. *Let one of the conditions*

a) $\alpha = \beta$, $k < 0$;

b) $\alpha > \beta$, $k \leq \dfrac{(\alpha-\beta)\sqrt{\alpha-\beta}}{\sqrt{2\beta(\alpha|z_0|-\beta l)}}$ *be valid. Then* $\theta \geq T_l$ *holds, where* θ *and* T_l *are given in (6.14), (6.21), respectively.*

Proof. a) Assume $\alpha = \beta$, $k < 0$. Then from $|z_0| > l$, we get

$$T_l = -\frac{1}{k}\left(1 - \frac{l}{|z_0|}\right) < -\frac{1}{k}.$$

b) Assume $\alpha > \beta$, $k \leq \dfrac{(\alpha-\beta)\sqrt{\alpha-\beta}}{\sqrt{2\beta(\alpha|z_0|-\beta l)}}$. Separate this case into two parts.

1) Let $\alpha > \beta$, $k \leq 0$. Then it is evident that the following is true:

$$(\alpha-\beta)^2 > k\beta\left(k|z_0| + \sqrt{|z_0|^2k^2 + 2(\alpha-\beta)(|z_0|-l)}\right). \tag{6.22}$$

Below transform (6.22) successively, i.e.,

$$2l\alpha(\alpha-\beta)^2 - 2l\alpha\beta k^2|z_0| > 2lk\beta\sqrt{\alpha^2k^2|z_0|^2 + 2\alpha^2(\alpha-\beta)(|z_0|-l)},$$

or

$$2\alpha(\alpha-\beta)(\alpha|z_0| - \beta l - \alpha|z_0| + \alpha l) - 2l\alpha\beta k^2|z_0| >$$
$$> 2lk\beta\sqrt{\alpha^2k^2|z_0|^2 + 2\alpha^2(\alpha-\beta)(|z_0|-l)}.$$

If we open the brackets on the left-hand side of the last inequality and add $\alpha^2k^2|z_0|^2 + l^2k^2\beta^2$ to both sides, then we get the following transformation:

$$\sqrt{k^2(\alpha|z_0|-\beta l)^2 + 2\alpha(\alpha-\beta)(\alpha|z_0|-\beta l)} >$$
$$> \sqrt{\alpha^2k^2|z_0|^2 + 2\alpha^2(\alpha-\beta)(|z_0|-l)} + lk\beta.$$

Let us add $\alpha|z_0|k$ to both sides of the final inequality and multiply it by $1/(\alpha^2-\alpha\beta)$. Then we obtain $\theta > T_l$.

2) Let $\alpha > \beta$, $0 < k \leq \dfrac{(\alpha-\beta)\sqrt{\alpha-\beta}}{\sqrt{2\beta(\alpha|z_0|-\beta l)}}$. Then from the second inequality it follows that

$$k^2 \leq \frac{(\alpha-\beta)^3}{2\beta(\alpha|z_0|-\beta l)}. \tag{6.23}$$

In (6.23), perform transformations as above, i.e.

$$2k^2\beta^2(|z_0|-l)+2k^2\beta|z_0|(\alpha-\beta)\le(\alpha-\beta)^3,$$

or

$$2k^2\beta^2(\alpha-\beta)(|z_0|-l)+2k^2\beta|z_0|(\alpha-\beta)^2\le(\alpha-\beta)^4.$$

Let's square both sides and let's turn to it

$$k\beta\sqrt{k^2|z_0|^2+2(\alpha-\beta)(|z_0|-l)}\le\left|(\alpha-\beta)^2-k^2\beta|z_0|\right|. \tag{6.24}$$

By virtue of (6.23), for the right-hand side of (6.24) we obtain

$$(\alpha-\beta)^2-k^2\beta|z_0|\ge(\alpha-\beta)^2\left(\frac{1}{2}+\frac{\beta(|z_0|-l)}{2(\alpha|z_0|-\beta l)}\right)>0.$$

Hence, we can omit the module in (6.24), i.e.,

$$k\beta\sqrt{k^2|z_0|^2+2(\alpha-\beta)(|z_0|-l)}\le(\alpha-\beta)^2-k^2\beta|z_0|,$$

or

$$k\beta\left(k|z_0|+\sqrt{|z_0|^2k^2+2(\alpha-\beta)(|z_0|-l)}\right)\le(\alpha-\beta)^2. \tag{6.25}$$

We can see that for the case $0<k\le\dfrac{(\alpha-\beta)\sqrt{\alpha-\beta}}{\sqrt{2\beta(\alpha|z_0|-\beta l)}}$ the inequality (6.25) holds. By applying the sequence of transformations performed for (6.22) to (6.25), it appears that the relation $T_l\le\theta$ is evident. The proof is completed. □

Definition 3.2. *We say that the Π_{G_l}-strategy (6.10) guarantees to occur l-capture on time interval $[0,T_l]$ if for any control $v(\cdot)\in\mathscr{V}_G$ of the evader*

a) there exists some time moment t^*, $t^*\in[0,T_l]$ at which the solution $z(t)$ of the equation

$$\ddot{z}=\mathbf{u}^*_{G_l}(z_0,v(t))-v(t),\ \ z(0)=z_0,\ \ \dot{z}(0)=z_1$$

satisfies the inequality $|z(t^*)|\le l$;

b) an inclusion $\mathbf{u}^*_{G_l}(z_0,v(\cdot))\in\mathscr{U}_G$ is satisfied on the time interval $[0,t^*]$.

Here, we call the number T_l a guaranteed time of l-capture in the G_l-Game.

Theorem 3.5. *Let the condition of Lemma 3.4 be satisfied. Then the Π_{G_l}-strategy (6.10) guarantees to occur l-capture on the time interval $[0,T_l]$, where T_l is the same with (6.21).*

Proof. Assume the evader chooses some control $v(\cdot) \in \mathscr{V}_G$, while the pursuer implements the Π_{G_l}-strategy (6.10). Then, by virtue of (6.12), we obtain the function

$$z(t) = z_0(1+kt) - \int_0^t (t-s)\lambda(z_0, v(s)) \frac{\alpha z_0 + v(s)l}{\alpha + \lambda(z_0, v(s))l} ds. \tag{6.26}$$

Due to (6.13), we can express (6.26) as

$$z(t) = z_0 \Lambda(t, v(\cdot)) + \int_0^t (t-s)\lambda(z_0, v(s)) \frac{v(s)l}{\alpha + \lambda(z_0, v(s))l} ds.$$

Since $\Lambda(t, v(\cdot)) \geq 0$ is valid in the interval $[0, \theta]$ (according to Lemma 3.3) and taking into account (6.4), we perform the following estimations for the absolute value of $z(t)$:

$$|z(t)| \leq |z_0| \Lambda(t, v(\cdot)) + \int_0^t (t-s)\lambda(z_0, v(s)) \frac{|v(s)|l}{\alpha + \lambda(z_0, v(s))l} ds,$$

or

$$|z(t)| \leq |z_0| \Lambda(t, v(\cdot)) + l\beta \int_0^t (t-s) \frac{\lambda(z_0, v(s))}{\alpha + \lambda(z_0, v(s))l} ds. \tag{6.27}$$

Substitute (6.13) into (6.27) and we get

$$|z(t)| \leq |z_0|(1+kt) - (\alpha|z_0| - \beta l) \int_0^t (t-s) \frac{\lambda(z_0, v(s))}{\alpha + \lambda(z_0, v(s))l} ds. \tag{6.28}$$

Apply (6.18) to the integral in (6.28) and reduce to the simple form

$$|z(t)| \leq |z_0|(1+kt) - \frac{\alpha - \beta}{2} t^2. \tag{6.29}$$

According to (6.20), we can write (6.29) down as

$$|z(t)| - l \leq F(t), \tag{6.30}$$

where $F(t)$ is identical with (6.20). In accordance with an examination of function $F(t)$, we have $F(T_l) = 0$ and from (6.30), assert that $|z(T_l)| \leq l$ or $|x(T_l) - y(T_l)| \leq l$. The proof of Theorem 3.5 is complete. $\square$

4. The I_l-Game

In this section, we will consider l-capture problem with inertial objects whose motion dynamics are given by the equations (6.1)-(6.2) and controls are picked out from the L_2 class. This game has been described as the I_l-Game above. To solve l-capture problem, we propose an approach strategy for the pursuer and obtain sufficient conditions for a solution. As we mentioned in the precious section, to achieve the main goal of object P, it is insufficient to rely solely on program strategies that depend exclusively on time t. Consequently, we can define the strategy for object P by basing it solely on the current positions of the acceleration function $v(t)$ for $t \geq 0$, along with the specified constants l, z_0, ρ, σ.

Remark 4.1. *We consider the I_l-Game for the case $x_1 = y_1$.*

Now we will define an approach strategy for the pursuer according to Krasovsky-Pontryagin's formalization and Pshenichnyi-Chikrii's method of resolving functions.

Definition 4.1. *For $\rho > \sigma$ we call the function*

$$\mathbf{u}^*_{I_l}(z_0, v) = v + \gamma(z_0, v)\,(m(z_0, v) - z_0) \tag{6.31}$$

the approach strategy or the Π_{I_l}-strategy for the pursuer, where

$$\gamma(z_0, v) = \max\left\{0, \frac{1}{h^2}\left[h(\delta + 2\langle v, z_0\rangle) + 2l^2\delta + 2l|z_0\delta + hv|\right]\right\}, \tag{6.32}$$

$$m(z_0, v) = -\frac{v - \gamma z_0}{|v - \gamma z_0|}l,\ h = |z_0|^2 - l^2,\ \delta = \rho^2 - \sigma^2,$$

and $\langle v, z_0\rangle$ denotes the scalar product of the vectors v and z_0 in $\mathbf{R}^n$.

Note that the function $\gamma(z_0, v)$ in (6.32) is generally called *the resolving function.*

Proposition 4.2. *If $\rho > \sigma$, then the function $\mathbf{u}^*_{I_l}(z_0, v)$ in (6.31) is defined, continuous and the equality*

$$|\mathbf{u}^*_{I_l}(z_0, v)|^2 = |v|^2 + \delta\gamma(z_0, v)$$

holds during the l-capture problem in the I_l-Game.

Consider the equation

$$1 - \int_0^t (t - s)\gamma(z_0, v(s))ds = 0 \tag{6.33}$$

with respect to $t \geq 0$. Let $\Omega(t, v(\cdot)) = 1 - \int\limits_0^t (t-s)\gamma(z_0, v(s))ds$ and let $T = T(z_0, v(\cdot))$ be the first positive root of the equation (6.33) for an optional admissible control $v(\cdot)$ of the evader.

Definition 4.2. *We say that the Π_{I_l}-strategy (6.31) guarantees to occur l-capture on time interval $[0, \eta]$ if for any control $v(\cdot) \in \mathscr{V}_I$ of the Evader*

a) there exists some time moment T, $T \in [0, \eta]$ at which the solution $z(t)$ of the equation

$$\ddot{z} = \mathbf{u}^*_{I_l}(z_0, v(t)) - v(t), \;\; z(0) = z_0, \;\; \dot{z}(0) = 0 \tag{6.34}$$

satisfies the inequality $|z(T)| \leq l$;

b) an inclusion $\mathbf{u}^*_{I_l}(z_0, v(\cdot)) \in \mathscr{U}_I$ is satisfied on the time interval $[0, T]$.

Here we call the time η a guaranteed time of l-capture in the I_l-Game and

$$\eta = \frac{|z_0| - l}{\rho - \sigma}\sqrt{2}.$$

The existence and boundedness of the root $T(z_0, v(\cdot))$ will be shown in the following statement.

Theorem 4.3. *Let $\rho > \sigma$ be valid in the I_l-Game. Then from the point z_0, $|z_0| > l$, the Π_{I_l}-strategy (6.31) guarantees to occur l-capture on the time interval $[0, T(z_0, v(\cdot))]$.*

Proof. Let $v(\cdot)$ be an optional admissible control of E. First of all, we will show the existence and boundedness of the root $T(z_0, v(\cdot))$ of the equation (6.33). For this purpose, we may write the following estimates for the function $\Omega(t, v(\cdot))$:

$$\Omega(t, v(\cdot)) = 1 - \int\limits_0^t (t-s)\gamma(z_0, v(s))ds =$$

$$1 - \frac{1}{h^2}\int\limits_0^t (t-s)\max\left\{0, h(\delta + 2\langle v(s), z_0\rangle) + 2l^2\delta + 2l|z_0\delta + v(s)h|\right\}ds \leq$$

$$\leq 1 - \frac{1}{h^2}\max\left\{0, \Gamma(t, v(\cdot))\right\},$$

or

$$\Omega(t, v(\cdot)) \leq 1 - \frac{1}{h^2}\max\left\{0, \Gamma(t, v(\cdot))\right\}, \tag{6.35}$$

where

$$\Gamma(t,v(\cdot)) = \frac{t^2\delta}{2}(h+2l^2) + 2h\int_0^t (t-s)\langle v(s), z_0\rangle ds +$$

$$+2l\int_0^t (t-s)|z_0\delta + v(s)h|ds.$$

Now, using the property of the absolute value of subtraction of two vectors, estimate only the function $\Gamma(t,v(\cdot))$, that is,

$$\Gamma(t,v(\cdot)) \geq \frac{t^2\delta}{2}(h+2l^2) -$$

$$-2h\int_0^t (t-s)|v(s)||z_0|ds + 2l\int_0^t (t-s)(|z_0|\delta - |v(s)|h)ds =$$

$$= \frac{t^2\delta}{2}(h+2l^2+2l|z_0|) - (2h|z_0|+2lh)\int_0^t (t-s)|v(s)|ds.$$

Applying the Cauchy–Bunyakovsky inequality to the integral in the last expression and taking account of the constraint (6.6), we obtain the following:

$$\int_0^t (t-s)|v(s)|ds \leq \sqrt{\int_0^t (t-s)ds}\sqrt{\int_0^t (t-s)|v(s)|^2 ds} \leq \frac{t\sigma}{\sqrt{2}}.$$

From this we generate the final estimate for the function $\Gamma(t,v(\cdot))$, that is,

$$\Gamma(t,v(\cdot)) \geq \frac{t^2\delta}{2}(|z_0|+l)^2 - \sqrt{2}ht\sigma(|z_0|+l) = \Gamma(t). \tag{6.36}$$

According to (6.35) and (6.36), we get the inequality

$$\Omega(t,v(\cdot)) \leq 1 - \frac{1}{h^2}\max\{0, \Gamma(t)\}. \tag{6.37}$$

It can be easily checked that the right-hand side of (6.37) vanishes at time $t=\eta$, where

$$\eta = \frac{|z_0|-l}{\rho-\sigma}\sqrt{2}.$$

Consequently, due to the continuity of $\Omega(t,v(\cdot))$ with respect to t, $t \geq 0$, we have the inequality $\Omega(\eta, v(\cdot)) \leq 0$. From this and from $\Omega(0,v(\cdot)) = 1$, we can assert that there exists a moment $T = T(z_0, v(\cdot))$ such that $\Omega(T, v(\cdot)) = 0$. This implies that $T(z_0, v(\cdot)) \leq \eta$.

Let's now show that from an arbitrary point z_0, $|z_0| > l$, the l-capture ends exactly at the time $T(z_0, v(\cdot))$. To do this, for an arbitrary admissible control $v = v(t)$, $0 \leq t \leq T$, the pursuer implements the Π_{I_l}-strategy (6.31), i.e. the control $\mathbf{u}^*_{I_l}(z_0, v(t))$, $0 \leq t \leq T$.

Then we reduce the equation (6.34) to the form

$$\ddot{z} = \gamma(z_0, v(t))\,(m(z_0, v(t)) - z_0),$$

$$z(0) = z_0, \;\; \dot{z}(0) = 0, \;\; 0 \leq t \leq T.$$

Integrating both sides of the last equation in the interval $[0, T]$, we obtain the solution

$$z(T) = z_0 + \int\limits_0^T (T-s)\gamma(z_0, v(s))(m(z_0, v(s)) - z_0)ds.$$

Considering $\Omega(T, v(\cdot)) = 0$ in (6.33) and taking into account the form of the function $m(z_0, v(t))$, we get

$$|z(T)| - l =$$

$$= \left| z_0 - z_0 \int\limits_0^T (T-s)\lambda(z_0, v(s))ds + \int\limits_0^T (T-s)\lambda(z_0, v(s))m(z_0, v(s))ds \right| - l \leq$$

$$\leq \left(|z_0| - l\right)\left(1 - \int\limits_0^T (T-s)\lambda(z_0, v(s))ds\right),$$

or $|z(T)| \leq l$.

Admissibility of realization of the Π_{I_l}-strategy (6.31) follows from the constraint (6.6) and Proposition 4.2, and from the fact that T is a root of the equation (6.33). Firstly, integrate both sides of the equality

$$|\mathbf{u}^*_{I_l}(z_0, v(t))|^2 = |v(t)|^2 + \delta\gamma(z_0, v(t))$$

in the time interval $[0, T]$, i.e.

$$\int\limits_0^T (T-s)|\mathbf{u}^*_{I_l}(z_0, v(s))|^2 ds = \int\limits_0^T (T-s)|v(s)|^2 ds +$$

$$+\delta\int_0^T (T-s)\gamma(z_0, v(s))ds \le \sigma^2+\delta=\rho^2.$$

The proof of Theorem 4.3 has been completed. □

5. Conclusion

In this chapter, we considered l-capture problems for the differential game of one pursuer and one evader with the inertial movements. Object control parameters of the objects are selected from two different classes of measurable functions, and the game divided into two parts. In the first part, we imposed geometric constraints, and in the second part we imposed integral constraints on the object controls. It can be said that in both games the approach strategies constructed for the pursuer are effective in minimizing the occurrence of l-catch. The obtained sufficient conditions allow us to find the shortest guaranteed l-capture time. The following problems can be further explored:

- Problem of l-capture in a differential game with many pursuers and one evader under integral restrictions on the controls.
- Problem of l-capture in a differential game with one pursuer and one evader under integral-geometric restrictions on controls.

Bibliography

[1] Cruz-Uribe D.V., Fiorenza A. *Variable Lebesgue Spaces*. Foundation and Harmonic Analysis. Birkhäsuser, 2013.

[2] Adams D. *Morrey Spaces*. Lecture Notes in Applied and Numerical Harmonic Analysis, Birkhuser/Springer, xvi, 121 p., 2015.

[3] Kokilashvili V., Meskhi A., Rafeiro H., Samko S. *Integral Operators in Non-Standard Function Spaces*. Volume 1: Variable Exponent Lebesgue and Amalgam Spaces, Springer, 2016.

[4] Kokilashvili V., Meskhi A., Rafeiro H., Samko S. *Integral Operators in Non-Standard Function Spaces, Volume 2: Variable Exponent Holder, Morrey-Campanato, and Grand Spaces,* Operator Theory: Advances and Applications, **249**, Springer, 2016.

[5] Sawano Y., Fazio G.D., Hakim D.I. *Morrey Spaces*, Introduction and Applications to Integral Operators and PDE's, Volume I, 2016.

[6] Sawano Y., Fazio G.D., Hakim D.I. *Morrey Spaces*, Introduction and Applications to Integral Operators and PDE's, Volume II, 2020.

[7] Caso L., D'Ambrosio R., Softova L. *Generalized Morrey spaces over unbounded domains*, Azerbaijan Journal of Mathematics, **10(1)**, 2020, 193-208.

[8] Sharapudinov I.I. *On Direct and Inverse Theorems of Approximation Theory In Variable Lebesgue Space And Sobolev Spaces*, Azerbaijan Journal of Mathematics, **4(1)**, 2014, 55-72.

[9] Bilalov B.T., Ahmadov T.M., Zeren Y., Sadigova S.R. *Solution in the small and interior Schauder type Estimate for the m-th order Elliptic operator in Morrey- Sobolev spaces*, Azerbaijan Journal of Mathematics, **12(2)**, 2022, 190-219.

[10] Bilalov B.T. Sadigova S.R. *On local solvability of higher order elliptic equations in rearrangement invariant spaces*, Siberian Mathematical Journal, **63(3)**, 2022, 425-437

[11] Bilalov B.T., Sadigova S.R. *On the Fredholmness of the Dirichlet problem for a second-order elliptic equation in grand-Sobolev spaces*, Ricerche di Matematica, **73**, 2024, 283322.

[12] Bilalov B.T., Sadigova S.R., Sezer Y., Nasibova N.P. *On solvability of polyharmonic Dirichlet problem in symmetric Sobolev spaces*, Math. Meth. Appl. Sci., 2024, 116. DOI: 10.1002/mma.10279

[13] Bilalov B.T., Huseynli A.A., El-Shabrawy S.R. *Basis Properties of Trigonometric Systems in Weighted Morrey Spaces*, Azerbaijan Journal of Mathematics, **9(2)** , 2019, 200-226.

[14] Acquistapace P. *On BMO regularity for linear elliptic systems*, Ann. Mat. Pura Appl., **161**, 1992, 231-270.

[15] Adams R. *Sobolev Spaces.* Academic Press, New York, 1975.

[16] Almeida A., Samko S. *Approximation in Morrey spaces.* J. Funct. Anal., **72(6)**, 2017, 2392-2411.

[17] Almeida A., Samko S. *Approximation in generalized Morrey spaces*, Georgian Math. J., **25(2)**, 2018, 155-168.

[18] Brudny Yu A. *Local best approximation of functions by polynomials*, Sov. Math., Dokl., **6**, 1965, 483-486.

[19] Burenkov V.I., Gogatishvili A., Guliyev V.S., Mustafayev R. *Boundedness of the fractional maximal operator in local Morrey-type spaces*, Complex Var. Elliptic Equ., **55(8-10)**, 2010, 739-758.

[20] Campanato S. *Proprieta di inclusione per spazi di Morrey*, Ricerca Matematica, **12**, 1963, 67-86.

[21] Campanato S. *Proprietà di Hölderianità di alcune classi di funzioni*, Ann. Scuola Norm. Sup. Pisa, **17**, 1964, 175-188.

[22] Campanato S. *Proprietà di una famiglia di spazi funzionali*, Ann. Scuola Norm. Sup. Pisa, **18**, 1964, 137-160.

[23] Cavaliere P., Cianchi A., Pick L., Slavkov L. *Higher-order Sobolev embeddings into spaces of Campanato and Morrey type*, 2024. arXiv:2404.09702.

[24] Chiarenza F, Franciosi M. A generalization of a theorem by C. Miranda, *Ann. Mat. Pura Appl.* (1992) **161**(4):285-297.

[25] Chiarenza F, Frasca M. *Morrey spaces and Hardy-Littlewood maximal function*, Rend. Matematica, **7**, 1987, 273-279.

[26] Chiarenza F., Frasca M., Longo P. I*nterior $W^{2,p}$-estimates for non-divergence elliptic equations with discontinuous coefficients*, Ricerche Matem., **40**, 1991, 149-168.

[27] Chiarenza F., Frasca M., Longo P. *$W^{2,p}$-solvability of Dirichlet problem for nondivergence ellipic equations with VMO coefficients*, Trans. Amer. Mathem. Society, **336**, 1993, 841-853.

[28] Cianchi A., Pick L. *Sobolev embeddings into spaces of Campanato, Morrey, and Hlder type*, J. Math. Analysis Appl., **282(1)**, 2003, 128-150.

[29] Coifman R.R., Rochberg R. *Another characterization of BMO*, Proc. Amer. Math. Society, **79(2)**, 1980, 249-254.

[30] Di Fazio G., Palagachev D.K., Ragusa M.A. *Global Morrey regularity of strong solutions to the Dirichlet problem for elliptic equations with discontinuous coefficients*, J. Funct. Anal., **166(2)**, 1999, 179-196.

[31] Fefferman Ch., Stein El M. *Some maximal inequalities*, American J. Math., **93**, 1971, 107-115.

[32] Fabes E., Rivière N. *Singular integrals with mixed homogeneity*, Studia Mathem., **27**, 1966, 19-38.

[33] Garcia-Cuerva J., Rubio de Francia J. *Weighted Norm Inequalities and Related Topics*. North-Holland Mathematics Studies, **116**, North-Holland Publishing Co., Amsterdam, 1985.

[34] Guliyev V.S. *Boundedness of the maximal, potential and singular operators in the generalized Morrey spaces*, J. Inequal. Appl., **2009**, 2009, Art. ID 503948:20.

[35] Guliyev V.S., Softova L. *Global regularity in generalized Morrey spaces of solutions to nondivergence elliptic equations with VMO coefficients*, Potent. Analysis, **38**, 2013, 843-862.

[36] Guliyev V.S., Softova L. *Generalized Morrey regularity for parabolic equations with discontinuous data* Proc. Edinburgh Math. Society, II **58(1)**, 2015, 1-23.

[37] Guliyev V.S., Softova L. *Generalized Morrey estimates for the gradient of divergence form parabolic operators with discontinuous coefficients*, J. Diff. Equations, **259**, 2015, 2368-2387.

[38] Guliyev V.S., Aliyev S.S., Karaman T., Shukurov P. *Boundedness of sublinear operators and commutators on generalized Morrey spaces*, Integral Equ. Operator Theory, **71(3)**, 2011, 327-355.

[39] Janson S., Taibleson M., Weiss G. *Elementary characterizations of the Morrey-Campanato spaces*, In: Harmonic Analysis (Cortona, 1982), Lecture Notes in Mathematics Springer, Berlin, **992**, 1983, 101-114.

[40] John F., Nirenberg L. *On functions of bounded mean oscillation*, Commun. Pure Appl. Math., **14**, 1961, 415-426.

[41] Jones P.W. *Extension theorems for BMO*, Indiana Univ. Math. J., **29**, 1980, 41-66.

[42] Komori Y, Shirai S. *Weighted Morrey spaces and a singular integral operators*, Math. Nachrichten, **282(2)**, 2009, 219-231.

[43] Kronz M. *Some function spaces on spaces of homogeneous type*, Manuscr. Mathematica, **106**, 2001, 219-248.

[44] Kufner A., John O., Fuk S. *Function Spaces*. Monogr. Textb. Mech. Solids and Fluids, Noordhoff Intern. Publ. Leyden; Academia Prague **XV**, 1977.

[45] Meyers N.G. *Mean oscillation over cubes and Hölder continuity*, Proceed. Amer. Math. Society, **15**, 1964, 717-721.

[46] Mizuhara T. *Boundedness of some classical operators on generalized Morrey spaces*, Harmonic Analysis. Proc. Conf., Sendai/Jap. 1990, ICM-90 Satell. Conf. Proc., 1991, 183-189.

[47] Mizuhara T. *Relations between Morrey and Campanato spaces with some growth functions*, II. Proc. Harmon. Anal. Seminar, **11**, 1995, 67-74.

[48] Morrey C.B. *On the solutions of quasi-linear elliptic partial differential equations*, Trans. Amer. Math. Society, **43**, 1938, 126-166.

[49] Morrey C.B. *Second order elliptic equations in several variables and Hölder continuity*, Math. Zeitschrift, **72**, 1959, 146-164.

[50] Muckenhoupt B. *Weighted norm inequalities for the Hardy maximal function*, Trans. Amer. Math. Society, **165**, 1972, 207-226.

[51] Nakai E. *Hardy-Littlewood maximal operator, singular integral operators and the Riesz potentials on generalized Morrey spaces*. Math. Nachrichten, **166**, 1994, 95-103.

[52] Nakai E. *Pointwise multipliers on weighted BMO spaces*, Stud. Mathematica, **125(1)**, 1997, 35-56.

[53] Nakai E. *The Campanato, Morrey and Hölder spaces on spaces of homogeneous type*, Stud. Mathematica, **176(1)**, 2006, 1-19.

[54] Palagachev D.K., Softova L. *Singular integral operators, Morrey spaces and fine regularity of solutions to PDEs*, Pot. Analysis, **20**, 2004, 237-263.

[55] Palagachev D.K., Softova L. *A priori estimates and precise regularity for parabolic systems with discontinuous data*, Discrete Contin. Dyn. Systems, **13(3)**, 2005, 721-742.

[56] Palagachev D.K., Softova L. *Fine regularity for elliptic systems with discontinuous ingredients*, Arch. Mathematics, **86(2)**, 2006, 145-153.

[57] Peetre J. *On the theory of $\mathscr{L}_{p,\lambda}$ spaces*, J. Funct. Analysis, **4**, 1969, 71-87.

[58] Piccinini L.C. *Inclusioni tra spazi di Morrey*, Boll. Uni. Matem. Italiana, IV, **2**, 1969 95-99.

[59] Rafeiro H., Samko N., Samko S. *Morrey-Campanato spaces: an overview*, In: Operator Theory: Advances and Applications, Springer Basel AG, **228**, 2013, 293-323.

[60] Rokotoson M. *Equivalence between the growth of $\int_{B(x,r)} |\nabla u|^p\, dy$ and T in the equation $P[u] = T$*, J. Diff. Equations, **86(1)**, 1990, 102-122.

[61] Rosenthal M., Triebel H. *Morrey spaces, their duals and preduals*, Rev. Matem. Complutense, **28(1)**, 2015, 1-30.

[62] Sarason D. *Functions of vanishing mean oscillation*, Trans. Amer. Math. Society, **207**, 1975, 391-405.

[63] Sawano Y., Tanaka H. *Morrey spaces for non-doubling measures*, Acta Math. Sinica, Engl. Ser., **21(6)**, 2005, 1535-1544.

[64] Softova L. *Oblique derivative problem for parabolic operators with VMO coefficients*, Manuscr. Mathem., **103(2)**, 2000, 203-220.

[65] Softova L. *Parabolic equations with VMO coefficients in Morrey spaces*, Electr. J. Diff. Equ., **2001(51)**, 2001, 1-25.

[66] Softova L. *Singular integrals and commutators in generalized Morrey spaces*, Acta Math. Sin., Engl. Ser., 2006 **22(3)**, 2006, 757-766.

[67] Softova L. *Singular integral operators in Morrey spaces and interior regularity of solutions to systems of linear PDEs*, J. Glob. Optimiz., **40**, 2008, 427-442.

[68] Softova L. *Morrey-type regularity of solutions to parabolic problems with discontinuous data*, Manuscr. Mathematica, **136(3-4)**, 2011, 365-382.

[69] Softova L. *The Dirichlet problem for elliptic equations with VMO coefficients in generalized Morrey spaces*. in: Advances in Harmonic Analysis and Operator Theory, Springer Basel, **229**, 2013, 371–386.

[70] Softova L. *Parabolic oblique derivative problem with discontinuous coefficients in generalized Morrey spaces*, Ric. Matemat., **62**, 2013, 265-278.

[71] Spanne S. *Some function spaces defined using the mean oscillation over cubes*, Ann. Sc. Norm. Super. Pisa, Sci. Fis. Mat. III, **19**, 1965, 593-608.

[72] Spanne S. *Sur linterpolation entre les espaces* $L_k^{p\Phi}$, Ann. Sc. Norm. Super. Pisa, Sci. Fis. Mat., III, **20**, 1966, 625-648.

[73] Stein E. *Harmonic Analysis: Real-Variable Methods, Orthogonality and Oscillatory Integrals*, Princeton Univ. Press, Princeton 1993.

[74] Stein E.M, Wainger S. *Problems in harmonic analysis related to curvature*, Bull. Amer. Math. Society, **84(6)**, 1978, 1239-1295.

[75] Torchinsky A. *Real Variable Methods in Harmonic Analysis*. Pure and Applied Mathematics, **123**, Academic Press, Inc., Orlando, FL, 1986.

[76] Transirico M., Troisi M., Vitolo A. *Spaces of Morrey type and elliptic equations in divergence form on unbounded domains*, Boll. Un. Mat. Ital. B. VII, **9(1)**, 1995, 153-174.

[77] Vitanza C. *Functions with vanishing Morrey norm and elliptic partial differential equations*. In: Proc. Methods Real Anal. PDEs, Capri, Springer, 1990, 147-150.

[78] Zorko C. *Morrey spaces*, Proc. Amer. Math. Society, **98**, 1986, 586-592.

[79] Kuratowski K. *Sur les espaces complets*, Fundamenta Mathematicae, **1(15)**, 1930, 301–309.

[80] Darbo G. *Punti uniti in trasformazioni a codominio non compatto*, Rendiconti del Seminario Matematico della Universitá di Padova, **24**, 1955, 84–92.

[81] Aghajani A., Mursaleen M., Shole Haghighi A. *Fixed point theorems for Meir-Keeler condensing operators via measure of noncompactness*, Acta Mathematica Scientia, **35(3)**, 2015, 552–566.

[82] Banaś J., Mursaleen M. *Sequence Spaces and Measures of Noncompactness with Applications to Differential and Integral Equations*, Springer, 2014, 147–184.

[83] Keeler E., Meir A. *A theorem on contraction mappings*, J. Math. Anal. Appl, **28**, 1969, 326–329.

[84] Banaś J. *Measures of noncompactness in the space of continuous tempered functions*, Demonstratio Mathematica, **14(1)**, 1981, 127–134.

[85] Das A., Hazarika B., Arab R., Mursaleen M. *Solvability of the infinite system of integral equations in two variables in the sequence spaces c_0 and ℓ_1*, Journal of Computational and Applied Mathematics, **326**, 2017, 183–192.

[86] Malik I.A., Jalal T. *Infinite system of integral equations in two variables of hammerstein type in c_0 and ℓ_1 spaces*, Filomat, **33(11)**, 2019, 3441–3455.

[87] Samadi A., Avini M.M., Mursaleen M. *Solutions of an infinite system of integral equations of Volterra-Stieltjes type in the sequence spaces ℓ_p and c_0*, AIMS Math., **5(4)**, 2020, 3791–3808.

[88] Kreyszig E. *Introductory functional analysis with applications*. John Wiley & Sons, **17**, 1991.

[89] Ghasemi M., Khanehgir M., Allahyari R. *On Solutions of Infinite Systems of Integral Equations in N-variables in Spaces of Tempered Sequences c_0^β and ℓ_1^β* , Journal of Mathematical Analysis, **9(6)**, 2018, 1-16.

[90] Akgün R. *Trigonometric approximation of functions in generalized Lebesgue spaces with variable exponent*, Ukrainian Mathematical Journal, **63(1)**, 2011, 1–26.

[91] Akgun R. *Polynomial approximation of functions in weighted Lebesgue and Smirnov spaces with nonstandard growth*, Georgian Mathematical Journal, **18**, 2011, 203-235.

[92] Akgün R., Kokilashvili V. *On converse theorems of trigonometric approximation in weighted variable exponent Lebesgue spaces*, Banach Journal of Mathematical Analysis, **5(1)**, 2011, 70–82.

[93] Akgün R., Kokilashvili V. *Some notes on trigonometric approximation of (α, ψ)-differentiable functions in weighted variable exponent Lebesgue spaces*, Proceedings of A. Razmadze Mathematical Institute, **163**, 2013, 15–23.

[94] Alper S.Y. *Approximation in the mean of analytic functions of class Ep.* In book : Investigations on the modern problems of the function theory of a complex variable, M., Gos. Izdat. Fiz.-Mat. Lit., 1960, 272–286 (in Russian).

[95] Andersson J.E. *On the Degree of Polynomial Approximation in* $E^p(D)$, Journal of Approximation Theory, **19**, 1977, 61-68.

[96] Bilalov B.T., Mamedov F.I., Bandaliev R.A. *On classes of harmonic functions with variable exponent*, Dokl NAN Azerb, **63(5)**, 2007, 16-21.

[97] Bilalov B.T, Guseynov Z.G. *On basicity from exponents, sines and cosines in Lebesgue space with variable exponent*, The Proceeding of The 40-th Annual Iranian Mathemathics Conference, 2009, 439-444.

[98] Bilalov B.T. *On the basicity from exponents in Lebesgue spaces with variable exponent*, TWMS Journal of Pure and Applied Mathematics, **1(1)**, 2010, 14-23.

[99] Bilalov B.T., Guseinov Z.G. *A criterion for the basis property of perturbed exponential systems in Lebesgue spaces with variable exponent*, Doklady Mathematics, **83(1)**, 2011, 93-96.

[100] Cavus A., Israfilov D.M. *Approximation by Faber-Laurent rational functions in the mean of functions of the class* $L_p(\Gamma)$, *with* $1 < p < \infty$, Approximation Theory and its Applications, **11(1)**, 1995, 105-118.

[101] Chandra P. *Trigonometric approximation of functions in Lp-norm*, Journal of Mathematical Analysis and Applications, **275(1)**, 2002, 13-26.

[102] Cruz-Uribe D.V., Diening L., Hästö P. *The maximal operator on weighted variable Lebesgue spaces*, Fractional Calculus and Applied Analysis, **14(3)**, 2011, 361-374.

[103] Cruz-Uribe D.V., Wang D.L. *Extrapolation and weighted norm inequalities in the variable Lebesgue spaces*, Transactions of the American Mathematical Society, **369**, 2017, 1205-1235.

[104] De Vore R.A., Lorentz G.G. *Constructive Approximation : Polynomials and spline approximation*. Springer-Verlag, 1993.

[105] Diening L. *Maximal function on generalized Lebesgue spaces* $L^{p(x)}$, Mathematical Inequalities and Applications, **7(2)**, 2004, 245-253.

[106] Diening L., Harjulehto P., Hästö P., Růžiĉka M. *Lebesgue and Sobolev spaces with variable exponents*. Springer-Verlag, Berlin, 2011.

[107] Ditzian Z., Totik V. *Moduli of Smoothness*. Springer-Verlag, New York, 1987.

[108] Draganov B.R., Ivanov K.G. *A generalized modulus of smoothness*, Proceedings of the American Mathematical Society, **142(5)**, 2014, 1577-1590.

[109] Goluzin G.M. *Geometric Theory of Functions of a Complex Variable*, Translation of Mathematical Monographs, **26**, AMS, 1969.

[110] Guven A., Israfilov D.M. *Trigonometric Approximation in Generalized Lebesgue Spaces $L^{p(x)}$*, Journal of Mathematical Inequalities, **4(2)**, 2010, 285-299.

[111] Guven A. *Trigonometric approximation by by matrix transform in $L^{p(x)}$ space*, Analysis and Applications, **10(1)**, 2012, 47-65.

[112] Guven A. *Approximation in weighted L_p spaces*, Revista de la Union Matematica Argentina, **53(1)**, 2012, 11-23.

[113] Ibragimov I.I., Mamedkhanov D.I. *A constructive characterization of a certain class of functions*, Dokl. Akad. Nauk SSSR, **223**, 1975, 35-37; Soviet Math. Dokl., **4**, 1976, 820-823.

[114] Israfilov D.M. *Approximation by p− Faber polynomials in the weighted Smirnov class $E^p(G,\omega)$ and the Bieberbach polynomials*, Constructive Approximation, **17**, 2001, 335-351.

[115] Israfilov D.M., Guven A. *Approximation in weighted Smirnov Classes*, East Journal on Approximation, **11(1)**, 2005, 91-102.

[116] Israfilov D., Kokilashvili V., Samko S. *Approximation In Weighted Lebesgue and Smirnov Spaces With Variable Exponents*, Proceedings of A. Razmadze Mathematical Institute, **143**, 2007, 25-35.

[117] Israfilov D.M., Gürsel E., Aydın E. *Maximal convergence of Faber Series in Smirnov Classes with Variable Exponent*, Bulletin of the Brazilian Mathematical Society New Series, **49**, 2018, 955–963.

[118] Israfilov D.M., Gürsel E. *Faber–Laurent series in variable Smirnov classes*, Turkish Journal of Mathematics, **44(2)**, 2020, 389-402.

[119] Israfilov D.M., Gürsel E. *Direct and inverse theorems in variable exponent Smirnov classes*, Proceedings of the Institute of Mathematics and Mechanics, National Academy of Sciences of Azerbaijan, **47(1)**, 2021, 55–66.

[120] Israfilov D.M., Gürsel E. *Approximation by $p(\cdot)$-Faber polynomials in the variable Smirnov classes*, Mathematical Methods in the Applied Sciences, **44**, 2021, 7479–7490.

[121] Israfilov D.M., Testici A. *Approximation in Smirnov Classes with Variable Exponent*, Complex Variables and Elliptic Equations, **60(9)**, 2015, 1243-1253.

[122] Israfilov D.M., Testici A. *Approximation by Faber-Laurent rational functions in Lebesgue spaces with variable exponent*, Indagationes Mathematicae, **27(4)**, 2016, 914-922.

[123] Israfilov D.M., Testici A. *Approximation by Matrix Transforms in Weighted Lebesgue Spaces with Variable Exponent*, Results in Mathematics, **73(8)**, 2018. https://doi.org/10.1007/s00025-018-0762-4

[124] Israfilov D.M., Testici A. *Multiplier and Approximation Theorems in Smirnov Classes with variable exponent*, Turkish Journal of Mathematics, **42**, 2018, 1442-1456.

[125] Israfilov D.M., Testici A. *Some Inverse and Simultaneous Approximation Theorems in Weighted Variable Exponent Lebesgue Spaces*, Analysis Mathematica, **44(4)**, (2018), 475-492.

[126] Israfilov D.M., Testici A. *Simultaneous approximation in Lebesgue space with variable exponent*, Proceedings of the Institute of Mathematics and Mechanics ANAS Azerbaijan, **44(1)**, 2018, 3-18.

[127] Israfilov D.M., Testici A. *Approximation problems in the Lebesgue spaces with variable exponent*, Journal of Mathematical Analysis and Applications, **459(1)**, 2018, 112-123.

[128] Israfilov D.M., Testici A. *Approximation by Matrix Transforms in Generalized Grand Lebesgue Spaces with Variable exponent*, Applicable Analysis, **100(4)**, 2019, 819-834.

[129] Israfilov D.M., Testici A. *Linear methods of approximation in weighted Lebesgue spaces with variable exponent*, Hacettepe Journal of Mathematics and Statistics, **50(3)**, 2021, 744 – 753.

[130] Israfilov D.M., Testici A. *Matrix Transforms in Weighted Variable Exponent Lebesgue Space*, Azerbaijan Journal of Mathematics, **12(2)**, 2022, 30-44.

[131] Israfilov D.M., Yildirir Y. *Simultaneous and converse approximation theorems in weighted Lebesgue spaces*, Mathematical inequalities and applications, **14(2)**, 2011, 359-371.

[132] Jafarov S.Z. *Linear methods for summing Fourier series and approximation in weighted Lebesgue spaces with variable exponents*, Ukrainian Mathematical Journal, **66(10)**, 2015, 1348-1356.

[133] Jafarov S.Z. *Approximation by means of Fourier trigonometric series in weighted Lebesgue spaces*, Sarajevo Journals of Mathematics, **13(26)**, 2017, 217-226.

[134] Jafarov S.Z. *Approximation of the functions in weighted Lebesgue spaces with variable exponent*, Complex Variables and Elliptic Equations, **63(10)**, 2018, 1444-1458.

[135] Jafarov S.Z. *Approximation by Zygmund means in variable exponent Lebesque spaces*, Mathematica Moravica, **23(1)**, 2019), 27-39.

[136] Jafarov S.Z. *Approximation by means of Fourier series in Lebesgue spaces with variable exponent*, Kazakh Mathematical Journal, **3**, 2023), 57-68.

[137] Kasumov M.G. *On the basicity of Haar systems in the weighted variable exponent Lebesgue spaces*, Vladikavkaz Mathematical Journal, **16(3)**, 2014, 38-46.

[138] Kokilashvili V. *A direct theorem on Mean Approximation of Analytic Functions by Polynomials*, Soviet Mathematics Doklady, **10**, 1969, 411-414.

[139] Kokilashvili V., Samko S. *Weighted Boundedness In Lebesgue Spaces With Variable Exponents of Classical Operators on Carleson Curves*, Proceedings of A. Razmadze Mathematical Institute, **138**, 2005, 106-110.

[140] Kokilashvili V.M., Samko S.G. *Operators of harmonic analysis in weighted spaces with non-standard growth*, Journal of Mathematical Analysis and Applications, **352(1)**, 2009, 15-34.

[141] Leindler L. *Trigonometric approximation in Lp -norm*, Journal of Mathematical Analysis and Applications, **302(1)**, 2005, 129-136.

[142] Mittal B.E., Rhoades B.E., Mishra V.N., Uaday Singh. *Using infinite matrices to approximate functions of class image using trigonometric polynomials*, Journal of Mathematical Analysis and Applications, **326(1)**, 2007, 667-676.

[143] Mohapatra R.N., Russell D.C. *Some direct and inverse theorems in approximation of functions*, Journal of the Australian Mathematical Society (Ser. A), **34**, 1983, 143–154.

[144] Muckenhoupt B. *Weighted norm inequalities for the Hardy maximal function*, Transactions of the American Mathematical Society, **165**, 1972, 207-226.

[145] Orlicz W. *Über konjugierte Exponentenfolgen*, Studia Mathematica, **3**, 1931, 200-212.

[146] Potapov M.K., Berisha F.M. *Approximation of classes of functions defined by a generalized k-th modulus of smoothness*, East Journal on Approximation, **4(2)**, 1998, 217-241.

[147] Quade E.S. *Trigonometric approximation in the mean*, Duke Mathematical Journal, **3(3)**, 1937, 529–543; Informatica, **14(4)**, 422-427.

[148] Růžiĉka M. *Elektrorheological Fluids: Modeling and Mathematical Theory*. Lecture Notes in Mathematics, **1748**, Springer-Verlag, Berlin, 2000.

[149] Shakh-Emirov T.N. *On Uniform Boundedness of some Families of Integral Convolution Operators in Weighted Variable Exponent Lebesgue Spaces*, Izvestiya of Saratov University. Mathematics. Mechanics. Informatics, **14(4:1)**, 2014, 421-427.

[150] Sharapudinov I.I. *Some aspects of approximation theory in the spaces* $L^{p(x)}$, Analysis Mathematica, **33(2)**, 2007, 135-153.

[151] Sharapudinov I.I. *Some questions of approximation theory in the Lebesgue spaces with variable exponent*. Vladikavkaz, 2012.

[152] Sharapudinov I.I. *Approximation of functions in* $L_{2\pi}^{p(x)}$ *by trigonometric polynomials*, Izvestiya RAN : Ser. Math., **77(2)**, 2013, 197-224; English transl., Izvestiya : Mathematics, **77(2)**, 2013, 407-434.

[153] Smirnov V.I., Lebedev N.A. *Constructive theory of functions of complex variable*. M.- L.: Nauka, 1964 (in Russian).

[154] Suetin P. .*Series of Faber Polynomials*. Cordon and Breach Publishers, Moscow, Nauka, 1984.

[155] Timan A.F. *Theory of Approximation of Functions of a Real Variable*. New York, Macmillan, 1963.

[156] Testici A. *Approximation Theorems in Weighted Lebesgue Spaces with Variable Exponent*, Filomat, **35(2)**, 2021, 561–577.

[157] Volosivets S.S. *Realization functionals and description of a modulus of smoothness in variable exponent Lebesgue spaces*, Izvestiya Vysshikh Uchebnykh Zavedenii. Matematikai, **6(6)**, 2022, 13–25; Russian Math. (Iz. VUZ), **66(6)**, 2022, 8–19.

[158] Volosivets S.S. *Approximation of functions and their conjugates in variable Lebesgue spaces*, Matematicheskii Sbornik, **208(1)**, 2017, 48–64; Sb. Math., **208(1)**, 2017, 44–59.

[159] Volosivets S. *Approximation in Variable Exponent Spaces and Growth of Norms of Trigonometric Polynomials*, Analysis Mathematica, **49**, 2023, 307–336.

[160] Walsh J.L., Russel H.G., *Integrated continuity conditions and degree of approximation by polynomials or by bounded analytic functions*, Transactions of the American Mathematical Society, **92**, 1959, 355-370.

[161] Warschawski S. *Über das Randverhalten der Ableitung der Abbildungsfunktionen bei konformer Abbildung*, Mathematische Zeitschrift, **35**, 1932, 321-456.

[162] Yildirir Y.E., Israfilov D.M. *Approximation theorems in weighted Lorentz spaces*, Carpathian Journal of Mathematics, **26(1)**, 2010, 108-119.

[163] Zygmund A. *Trigonometric series*. Vols. I and II, Cambridge University Press, 1959)

[164] Askhabov S.N. *On a second-order integro-differential equation with difference kernels and power nonlinearity*, Bulletin of the Karaganda University. Mathematics series, **106(2)**, 2022, 38–48. DOI 10.31489/2022M2/38-48

[165] Assanova A.T. *A two-point boundary value problem for a fourth order partial integro-differential equation*, Lobachevskii Journal of Mathematics, **42(3)**, 2021, 526–535. https://doi.org/10.1134/S1995080221030082

[166] Assanova A.T., Nurmukanbet S.N. *A solvability of a problem for a Fredholm integro-differential equation with weakly singular kernel*, Lobachevskii Journal of Mathematics, **43(1)**, 2022, 182–191. https://doi.org/10.1134/S1995080222040047

[167] Aviltay N., Akhmet M. *Asymptotic behavior of the solution of the integral boundary value problem for singularly perturbed integro-differential equations*, Journal of Mathematics, Mechanics and Computer Science, **112(4)**, 2021, 13–28. https://doi.org/10.26577/JMMCS.2021.v112.i4.02

[168] Bellour A., Rouiban Kh. *Iterative continuous collacation method for solving nonlinear Volterra integro-differential equations*, Proceedings of the Institute of Mathematics and Mechanics. National Academy of Sciences of Azerbaijan, **47(1)**, 2021, 99–111. https://doi.org/10.30546/2409-4994.47.1.99

[169] Chistyakov V.F., Chistyakova E.V. *Properties of degenerate systems of linear integro-differential equations and initial value problems for these equations*, Differential Equations, **59(1)**, 2023, 13–28. https://doi.org/10.1134/S0012266123010023

[170] Durdiev D.K., Safarov J.S. *Finding the two-dimensional relaxation kernel of an integro-differential wave equation*, Differential Equations, **59(2)**, 2023, 214–229. https://doi.org/10.1134/S0012266123020064

[171] Iskandarov S. *Lower bounds for the solutions of a first-order linear homogeneous Volterra integro-differential equation*, Differential Equations, **31(9)**, 1995, 1462–1466.

[172] Iskandarov S., Khalilov G.T. *On lower estimates of solutions and their derivatives to a fourth-order linear integrodifferential Volterra equation*, J. Math. Sci. (N. Y.), **230(5)**, 2018, 688–694.

[173] Rautian N.A., Vlasov V.V. *Spectral analysis of the generators for semigroups associated with Volterra integro-differential equations*, Lobachevskii J. of Mathematics, **44(3)**, 2023, 926–935.

[174] Yurko V.A. *Inverse problems for first-order integro-differential operators* Math. Notes, **100(6)**, 2016, 876–882.

[175] Yuldashev T.K. *On a boundary value problem for a fifth order partial integro-differential equation*, Azerbaijan Journal of Mathematics, **12(2)**, 2022, 154–172.

[176] Yuldashev T.K., Fayziyev A.K. *Inverse problem for a second order impulsive system of integro-differential equations with two redefinition vectors and mixed maxima*, Nanosystems: Physics. Chemistry. Mathematics, **14(1)**, 2023, 13–21. https://doi.org/10.17586/2220-8054-2023-14-1-13-21

[177] Yuldashev T.K., Eshkuvatov Z.K., Nik Long N.M.A. *Nonlinear Fredholm functional-integral equation of first kind with degenerate kernel and integral maxima*, Malaysian Journal of Fundamental and Applied Sciences, **19**, 2023, 82–92.

[178] Vasilyev V.B., Kutaiba Sh.H. *Asymptotic behavior of solution to some boundary value problem*, Azerbaijan Journal of Mathematics, **13(1)**, 2023, 51–61.

[179] Zavizion G.V. *Asymptotic solutions of systems of linear degenerate integro-differential equations*, Ukr. Math. Journal, **55(4)**, 2003, 521–534.

[180] Smirnov Yu.G. *On the equivalence of the electromagnetic problem of diffraction by an inhomogeneous bounded dielectric body to a volume singular integro-differential equation*, Comput. Math. and Math. Physics, **56(9)**, 2016, 1631–1640.

[181] Yakar A., Kutlay H. *Extensions of some differential inequalities for fractional integro-differential equations via upper and lower solutions*, Bulletin of the Karaganda University. Mathematics series, **109(1)**, 2023, 156–167. https://doi.org/10.31489/2023M1/156-167

[182] Yuldashev T.K., Zarifzoda S.K. *On a new class of singular integro-differential equations*, Bulletin of the Karaganda University. Mathematics series, **101(1)**, 2021, 138–148. https://doi.org/10.31489/2021M1/138-148

[183] Boichuk A.A., Strakh A.P. *Noetherian boundary-value problems for systems of linear integro-dynamical equations with degenerate kernel on a time scale*, Nelineynye kolebaniya, **17(1)**, 2014, 32–38 (in Russian).

[184] Djumabaev D.S., Bakirova E.A. *On one single solvability of boundary value problem for a system of Fredholm integro-differential equations with degenerate kernel*, Nelineynye kolebaniya, **18(4)**, 2015, 489–506 (in Russian).

[185] Yuldashev T.K. *Inverse problem for a nonlinear Benney–Luke type integro-differential equations with degenerate kernel*, Russian Mathematics, **60(8)**, 2016, 53–60.

[186] Yuldashev T.K. *Nonlocal mixed-value problem for a Boussinesq-type integro-differential equation with degenerate kernel*, Ukrainian Mathematical J., **68(8)**, 2017, 1278–1296.

[187] Yuldashev T.K. *Mixed problem for pseudoparabolic integro-differential equation with degenerate kernel*, Differential equations, **53(1)**, 2017, 99–108.

[188] Yuldashev T.K. *On Fredholm partial integro-differential equation of the third order*, Russian Mathematics, **59(9)**, 2015, 62–66.

[189] Yuldashev T.K., Yu.P. Apakov, A.Kh. Zhuraev, *Boundary value problem for third order partial integro-differential equation with a degenerate kernel*, Lobachevskii J. of Mathematics, **42(6)**, 2021, 1317–1327.

[190] Yuldashev T.K. *Determination of the coefficient and boundary regime in boundary value problem for integro-differential equation with degenerate kernel*, Lobachevskii Journal of Mathematics, **38(3)**, 2017, 547–553.

[191] Yuldashev T.K. *Spectral features of the solving of a Fredholm homogeneous integro-differential equation with integral conditions and reflecting deviation*, Lobachevskii Journal of Mathematics, **40(12)**, 2019, 2116–2123. https://doi.org/10.1134/S1995080219120138

[192] Yuldashev T.K. *On inverse boundary value problem for a Fredholm integro-differential equation with degenerate kernel and spectral parameter*, Lobachevskii Journal of Mathematics, **40(2)**, 2019, 230–239. https://doi.org/10.1134/S199508021902015X

[193] Yuldashev T.K. *On the solvability of a boundary value problem for the ordinary Fredholm integrodifferential equation with a degenerate kernel*, Computational Mathematics and Math. Physics, **59(2)**, 2019, 241–252. https://doi.org/10.1134/S0965542519020167

[194] Artykova Zh.A., Bandaliyev R.A., Yuldashev T.K. *Nonlocal direct and inverse problems for a second order nonhomogeneous Fredholm integro-differential equation with two redefinition data*, Lobachevskii Journal of Mathematics, **44(10)**, 2023, 4215–4230. https://doi.org/10.1134/S1995080223100050

[195] Yuldashev T.K., Artykova Zh.A., Alladustov Sh.U. *Nonlocal problem for a second order Fredholm integro-differential equation with degenerate kernel and real parameters*, Proceedings of the Institute of Mathematics and Mechanics of National Academy of Sciences of Azerbaijan, **49(2)**, 2023, 228–242. https://doi.org/10.30546/2409-4994.2023.49.2.228

[196] Aizenberg L.A. *Carleman's formulas in complex analysis*. Nauka, Novosibirsk, 1990.

[197] Arbuzov E.V., Bukhgeim A.L. *The Carleman formula for the Helmholtz equation*, Sib. Math. Jour., **47(3)**, 2006, 518-526.

[198] Ahmad Sh., Ullah A., Ahmad Sh., Saifullah S., Shokri, A. *Periodic solitons of Davey Stewartson Kedomtsev Petviashvili equation in (4+1)-dimension*, Res. in Phys. **50**, 2023, 1-7.

[199] Bulnes J. *An unusual quantum entanglement consistent with Schrödinger's equation*, Glob. and Stoch. Anal., **9(2)**, 2022, 78-87.

[200] Bulnes J. *Solving the heat equation by solving an integro-differential equation*, Glob. and Stoch. Anal., **9(2)**, 2022, 89-97.

[201] Berdawood K., Nachaoui A., Saeed R., Nachaoui M., Aboud F. *An efficient D-N alternating algorithm for solving an inverse problem for Helmholtz equation*, Disc. & Cont. Dynam. Syst., **14**, 2021, 1-22.

[202] Carleman T. *Les fonctions quasi analytiques*. Gautier-Villars et Cie., Paris, 1926.

[203] Ivanov V.K. *About incorrectly posed tasks*, Math. Collect., **61**, 1963, 211-223.

[204] Elrawy A., Abul-Dahab M.A., Alshehri M., Boulaaras S., Saleem M.A. *Results on building fractional matrix differential equation systems using a class of block matrices*, Fract., **30(10)**, 2022, 1-9.

[205] Fayziev Yu., Buvaev Q., Juraev D., Nuralieva N. Sadullaeva Sh. *The inverse problem for determining the source function in the equation with the Riemann-Liouville fractional derivative*, Glob. and Stoch. Anal. **9(2)**, 2022, 43-52.

[206] Goluzin G.M., Krylov, V.M. *The generalized Carleman formula and its application to the analytic continuation of functions*, Sbor. Math., **40(2)**, 1933, 144-149.

[207] Hadamard J. *The Cauchy problem for linear partial differential equations of hyperbolic type*, Nauka, Moscow, 1978.

[208] Ibrahimov V.R., Mehdiyeva G.Yu., Yue X.G., Kaabar M.K.A., Noeiaghdam S., Juraev D.A. *Novel symmetric numerical methods for solving symmetric mathematical problems*, Int. Jour. of Cir., Sys. and Sig. Proc., **15**, 2021, 1545-1557.

[209] Ibrahimov V.R., Imanova M.N. *The new way to solve physical problems described by ODE of the second order with the special structure*, WSEAS Tran. on Sys., **22**, 2023, 199-206.

[210] Islamov B.I., Kholbekov J.A. *On a non-local boundary value problem for a loaded parabolo-hyperbolic equation with three lines of type change*, Bull. of the Sam. Stat. Tech. Uni. Ser. Phys. and Math. Sci., **25(3)**, 2021, 407-422.

[211] Javed S., Masood S. and Shokri A. *Generalized class of finite population variance in the presence of random nonresponse using simulation approach*, Comp., **2**, 2023, 1-16.

[212] Jwamer K.H.F., Hilmi H.D. *Asymptotic behavior of eigenvalues and eigenfunctions of T.Regge fractional problem*, Jour. of Al-Qad. for Com. Sci. and Math., **14(3)**, 2022, 89-100.

[213] Juraev D.A. *The construction of the fundamental solution of the Helmholtz equation*, Rep. of the Acad. of Sci. of the Rep. of Uz., **2**, 2012, 14-17.

[214] Juraev D.A. *Regularization of the Cauchy problem for systems of equations of elliptic type*, Saarbrucken: LAP Lambert Academic Publishing, Germany, 2014.

[215] Juraev D.A. *Regularization of the Cauchy problem for systems of elliptic type equations of first order*, Uz. Math. Jour., **2**, 2016, 61-71.

[216] Juraev D.A. *The Cauchy problem for matrix factorizations of the Helmholtz equation in an unbounded domain*, Sib. Elec. Math. Rep., **14**, 2017, 752-764.

[217] Juraev D.A. *On the Cauchy problem for matrix factorizations of the Helmholtz equation in a bounded domain*, Sib. Elec. Math. Rep., **15**, 2018, 11-20.

[218] Juraev D.A. *The Cauchy problem for matrix factorizations of the Helmholtz equation in* $\mathbb{R}^3$, Jour. of Uni. Math., **1(3)**, 2018, 312-319.

[219] Juraev D.A. *On the Cauchy problem for matrix factorizations of the Helmholtz equation in an unbounded domain in* $\mathbb{R}^2$, Sib. Elec. Math. Rep., **15**, 2018, 1865-1877.

[220] Juraev D.A. *On the Cauchy problem for matrix factorizations of the Helmholtz equation*, Jour. of Uni. Math., **2(2)**, 2020, 113-126.

[221] Juraev D.A. *The solution of the ill-posed Cauchy problem for matrix factorizations of the Helmholtz equation*, Adv. Math. Mod. & Appl., **5(2)**, 2020, 205-221.

[222] Juraev D.A., Noeiaghdam S. *Regularization of the ill-posed Cauchy problem for matrix factorizations of the Helmholtz equation on the plane*, Axi., **10(2)**, 2021, 1-14.

[223] Juraev D.A. *Solution of the ill-posed Cauchy problem for matrix factorizations of the Helmholtz equation on the plane*, Glob. and Stoch. Anal., **8(3)**, 2021, 1-17.

[224] Juraev D.A., Noeiaghdam S. *Modern problems of mathematical physics and their applications*, Axi., **11(2)**, 2022, 1-6.

[225] Juraev D.A., Gasimov Y.S. *On the regularization Cauchy problem for matrix factorizations of the Helmholtz equation in a multidimensional bounded domain*, Az. Jour. of Math., **12(1)**, 2022, 142-161.

[226] Juraev D.A., Noeiaghdam S. *Modern problems of mathematical physics and their applications*, Axioms, MDPI, Basel, Switzerland, 2022.

[227] Juraev D.A. *On the solution of the Cauchy problem for matrix factorizations of the Helmholtz equation in a multidimensional spatial domain*, Glob. and Stoch. Anal., **9(2)**, 2022, 1-17.

[228] Juraev D.A. *The solution of the ill-posed Cauchy problem for matrix factorizations of the Helmholtz equation in a multidimensional bounded domain*, Pal. Jour. of Math., **11(3)**, 2022, 604-613.

[229] Juraev D.A., Shokri A. Marian D. *Solution of the ill-posed Cauchy problem for systems of elliptic type of the first order*, Frac. and Frac., **6(3)**, 2022, 1-11.

[230] Juraev D.A., Shokri A. Marian D. *On an approximate solution of the Cauchy problem for systems of equations of elliptic type of the first order*, Entr., **24(7)**, 2022, 1-18.

[231] Juraev D.A., Shokri A. Marian D. *On the approximate solution of the Cauchy problem in a multidimensional unbounded domain*, Frac. and Frac., **6(7)**, 2022, 1-14.

[232] Juraev D.A., Shokri A., Marian D. *Regularized solution of the Cauchy problem in an unbounded domain*, Sym., **14(8)**, 2022, 1-16.

[233] Juraev D.A., Cavalcanti M.M. *Cauchy problem for matrix factorizations of the Helmholtz equation in the space* $\mathbb{R}^m$, Bol. da Soc. Par. de Mat., **41(3)**, 2023, 1-12.

[234] Juraev D.A. *The Cauchy problem for matrix factorization of the Helmholtz equation in a multidimensional unbounded domain*, Bol. da Soc. Par. de Mat., **41(3)**, 2023, 1-18.

[235] Juraev D.A., Ibrahimov V. Agarwal P. *Regularization of the Cauchy problem for matrix factorizations of the Helmholtz equation on a two-dimensional bounded domain*, Pal. Jour. of Math., **12(1)**, 2023, 381-403.

[236] Juraev D.A. *Fundamental solution for the Helmholtz equation*, Eng. Appl., **2(2)**, 2023, 164-175.

[237] Juraev D.A., Jalalov M.J., Ibrahimov V.R. *On the formulation of the Cauchy problem for matrix factorizations of the Helmholtz equation*, Eng. Appl., 2(2), 2023, 176-189.

[238] Juraev D.A., Agarwal P., Shokri A., Elsayed E.E., Bulnes J.D. *On the formulation of the Cauchy problem for matrix factorizations of the Helmholtz equation*, Eng. Appl., **2(2)**, 2023, 176-189.

[239] Kwari L.J., Sunday J., Ndam J.N., Shokri A. Wang Y. *On the simulations of second-order oscillatory problems with applications to physical systems*, Axi., **12(3)**, 2023, 1-20.

[240] Kythe P.K. *Fundamental solutions for differential operators and applications*, Boston, Birkhauser, 1996.

[241] Lavrent'ev M.M. *On the Cauchy problem for second-order linear elliptic equations*, Rep. of the USSR Acad. of Sci., **112(2)**, 1957, 195-197.

[242] Lavrent'ev M.M. *On some ill-posed problems of mathematical physics*, Nauka, Novosibirsk, 1962.

[243] Nuriyeva V. *On the construction bilateral multistep methods and its application to solve Volterra integral equation*, Int. Jour. of Eng. Res. in Comp. Sci. and Eng., **9(10)**, 2022, 61-62.

[244] Shokri A., Khalsaraei M.M., Noeiaghdam S., Juraev D.A. *A new divided difference interpolation method for twovariable functions*, Glob. and Stoch. Anal., **9(2)**, 2022, 19-26.

[245] Tarkhanov N.N. *Stability of the solutions of elliptic systems*, Funct. Anal. Appl., **19(3)**, 1985, 245-247.

[246] Tarkhanov N.N. *On the Carleman matrix for elliptic systems*, Rep. of the USSR Acad. of Sci., **284(2)**, 1985, 294-297.

[247] Tarkhanov N.N. *The Cauchy problem for solutions of elliptic equations*, **7**, Akad. Verl., Berlin, 1995.

[248] Tikhonov A.N. *On the solution of ill-posed problems and the method of regularization*, Rep. of the USSR Acad. of Sci., **151(3)**, 1963, 501-504.

[249] Yarmukhamedov Sh. *On the Cauchy problem for the Laplace equation*, Rep. of the USSR Acad. of Sci., **235(2)**, 1977, 281-283.

[250] Yarmukhamedov Sh. *On the extension of the solution of the Helmholtz equation*, Rep. of the Rus. Acad. of Sci., **357(3)**, 1997, 320-323.

[251] Zhuraev D.A., *Cauchy problem for matrix factorizations of the Helmholtz equation*, Ukr. Math. Jour., **69(10)**, 2018, 1583-1592.

[252] Alekseev V.M., Tikhomirov V.M., Fomin S.V. *Optimal Control.* Moscow, Nauka, 1979 (In Russian).

[253] Azamov A. *On the quality problem for simple pursuit games with constraint*, Serdica Bulgariacae math. Publ. Sofia, **12(3)**, 1986 38–43.

[254] Azamov A.A., Samatov B.T. *The Π-Strategy: Analogies and Applications*, The Fourth International Conference on Game Theory and Management, St. Petersburg, **4**, 2010, 33–47.

[255] Azimov A.Ya. *Linear differential pursuit game with integral constraints on the control*, Differ. Uravn., **11**, 1975, 1723–1731.

[256] Azimov A.Ya., Guseynov F.V. *Certain classes of differential games with integral constraints*, Izv. Akad. Nauk SSSR Tehn. Kibernet., **3**, 1972, 9–16.

[257] Berkovitz L.D. *Differential game of generalized pursuit and evasion*, SIAM J. Contr., **24(3)**, 1986 361–373. doi:10.1137/0324021

[258] Chikrii A.A. *Conflict-Controlled Processes*. Kluwer: Dordrecht, 1997. doi:10.1007/978-94-017-1135-7

[259] Fleming W.H. *The convergence problem for differential games*, II. Advances in Game Theory (2nd ed.), Annals of Math., **52**, 1964, 195-210. doi:10.1515/9781400882014

[260] Friedman A. *Differential Games* (2nd ed.), Pure and Applied Mathematics, New York: Wiley Interscience, **25**, 1971. doi:MR 0421700

[261] Grigorenko N.L. *Mathematical Methods of Control for Several Dynamic Processes*. Moscow, Izdat. Gos. Univ., 1990.

[262] Gusyatnikov P.B., Mokhon'ko E.Z. *On ℓ_∞-Escape in a Linear Many-Person Differential Game with Integral Constraints*, Prikl. Matem. Mekh., **44(4)**, 1980, 618–624.

[263] Ibragimov G.I., Salimi M. *Pursuit-Evasion Differential Game with Many Inertial Players* Mathematical Problems in Engineering, 2009, 1-15. doi:10.1155/2009/653723

[264] Isaacs R. *Differential games*. New York: John Wiley and Sons, 1965.

[265] Khaidarov B.K. *Positional l-catch in the game of one evader and several pursuers*, Prikl. Matem. Mekh., **48(4)**, 1984, 574–579.

[266] Krasovsky N.N., Subbotin A.I. *Game-Theoretical Control Problems*. New York, Springer, 1988.

[267] Nahin P.J. *Chases and Escapes: The Mathematics of Pursuit and Evasion*. Princeton, Princeton University Press, 2012.

[268] Petrosyan L.A., Dutkevich V.G. *Games with "a Survival Zone", Occasion L-catch*, Vestnic Leningrad State Univ., **13(3)**, 1969, 31-38.

[269] Petrosyan L.A. *Differential games of pursuit. Series on optimization.* Singapore: World Scientific Publishing, **2**, 1993).

[270] Petrov N.N. *"Soft" capture in Pontryagin's example with many participants*, J. Appl. Maths Mechs., **67(5)**, 2003, 671–680.

[271] Pontryagin L.S. *Selected Works*. Moscow, MAKS Press, 2014 (in Russian).

[272] Pshenichnii B.N. *Simple pursuit by several objects*, Cybernetics and System Analysis, **12(5)**, 1976, 484–485. doi:10.1007/BF01070036

[273] Subbotin A.I. *Generalization of the main equation of differential game theory*, Journal of Optimization Theory and Applications, **43(1)**, 1984, 103–133. doi:10.1007/BF00934749

[274] Subbotin A.I., Chentsov A.G. *Optimization of Guaranteed Result in Control Problems*. Moscow, Nauka, 1981 (in Russian).

[275] Samatov B.T. *The pursuit-evasion problem under integral-geometric constraints on pursuer controls*, Automation and Remote Control, **74(7)**, 2013, 1072–1081. doi:10.1134/S0005117913070023

[276] Samatov B.T. *Problems of group pursuit with integral constraints on controls of the players I* Cybernetics and Systems Analysis, **49(5)**, 2013, 756–767. doi:10.1007/s10559-013-9563-7

[277] Samatov B.T. *Problems of group pursuit with integral constraints on controls of the players II*, Cybernetics and Systems Analysis, **49(6)**, 2013, 907–921. doi:10.1007/s10559-013-9581-5

[278] Samatov B.T., Soyibboev U.B. *Applications of the* Π*-strategy when players move with acceleration, Proceedings of the IUTAM Symposium on Optimal Guidance and Control for Autonomous Systems*, **40(10)**, 2023, 167–181. doi:https://doi.org/10.1007/978-3-031-39303-7_10

[279] Samatov B.T., Uralova S.I. *Differential games with inertial players the Langenhop type constraints on controls*, Lobachevskii Journal of Mathematics, **44(10)**, 2023, 4370-4378. doi:10.1134/S1995080223100347

[280] Samatov B.T, Uralova S.I, Mirzamaxmudov U.A. *The problem of Ramchundra for a problem of l-capture*, Scientific Bulletin of Namangan State University, **1(2)**, 2019, 10–14. doi:https://uzjournals.edu.uz/namdu/vol1/iss2/2

[281] Satimov N.Yu. *Methods for Solving the Pursuit Problem in the Theory of Differential Games*. Tashkent, Izd-vo NUUz, 2003 (in Russian).

[282] Vagin D.A., Petrov N.N. *Pursuit of a group of evaders in the Pontryagin example*, Journal. Appl. Maths Mechs., **68**, 2004, 555–560.

Index